50 Years of Phytochemistry Research

Volume 43

Recent Advances in Phytochemistry

For further volumes:
http://www.springer.com/series/6474

David R. Gang
Editor

50 Years of Phytochemistry Research

Springer

Editor
David R. Gang
Institute of Biological Chemistry
Washington State University
Pullman
Washington
USA

ISBN 978-3-319-34523-9 ISBN 978-3-319-00581-2 (eBook)
DOI 10.1007/978-3-319-00581-2
Springer Cham Heidelberg New York Dordrecht London

Printed on acid-free paper

Springer is part of Springer Science+Business Media (www.springer.com)

Preface

Introduction to the 43rd Volume of the Recent Advances in Phytochemistry Series

This is the third volume since the reintroduction of the *Recent Advances in Phytochemistry* (*RAP*) series, an annual publication supported by the Phytochemical Society of North America. Topics appropriate for *RAP* include the biosynthesis of natural products and regulation of metabolism, the ecology of specialized metabolites and the evolution of their pathways, and the effects of natural products or plants on human health. Research appropriate for *RAP* involves genomics, proteomics, metabolomics, natural product structural determination and new technology development, medicinal chemistry and metabolic engineering, or any of the myriad of fields that are now closely associated with what may be called “traditional phytochemistry” and plant biochemistry. The advent of postgenomics-based ways of thinking, systems biology, synthetic biology, comparative genomics/proteomics/transcriptomics/metabolomics, and especially of the introduction and establishment of a mentality that leads to the support of large collaborative projects has opened up many new doors to scientists interested and versed in the (bio)chemistry of plants. The goal of *RAP* is to highlight these developments.

Two main types of articles are printed in *RAP*: Perspectives and Communications. Perspectives in *RAP* are expected to synthesize results from the primary literature and perhaps from new/novel results and place these in perspective relative to the broader field. These articles may be similar to review articles, but also are intended to present important ideas and hypotheses and may present proposals for interesting directions in the field. It is the hope of the Editorial Board that these articles will be of great value to a large audience. Communications are intended to present new advances in the field that will be of interest to a large audience. Articles of both types are typically solicited from the society membership based on the content of the annual meeting talks, but in keeping with the title “Recent Advances in Phytochemistry” the editorial board reserves the right to solicit additional Perspectives and/or Communications from non-attendees as well (e.g., where an editorial board member has knowledge of an interesting recent advancement that would be of general interest to the society membership).

All submissions to *RAP* go through a rigorous peer review process, overseen by the Editorial Board, which includes external review. *RAP* is indexed with Springer published journals. All *RAP* papers are available not only in the published volume form, but also electronically through Springer's online literature services. This marks a significant change from past volumes of *RAP* and it is the hope of the Editorial Board that this will lead to broader dissemination of the contents of and greater interest in *RAP*.

This 43rd volume of RAP includes a total of seven articles based on talks presented at the 50th Anniversary meeting of the PSNA, which was held at the Fairmont Orchid in Waikaloa, Hawai'i, USA. These seven Perspectives give a very good picture of the breadth of plant (bio)chemistry research in North America, which is also indicative of the state of the field worldwide. Each of these articles describes the integration of several different approaches to ask and then answer fascinating questions regarding the function of interesting plant metabolites, either in the plant itself or in interactions with the environment (natural setting or human health application).

Two perspectives outline very clearly the power of approaching biological questions from a modern "omics" or systems biology approach. Beale and Ward outline how metabolomics approaches can be brought to bear on plant biosynthetic questions and quickly lead to important advances in our understanding or how plants produce important metabolites. Zandkarimi et al. outline the integration of ion mobility spectrometry into mass spectrometry-based metabolomics investigations and show clearly how powerful those two spectrometric technologies can be when used together.

Plant biotechnology and its application to plant protection, pathogen/pest deterrence and drug production is discussed in three perspectives. Asano et al. describe the development of plant cell cultures and tissue culture techniques that lead to production of important indole alkaloid compound production, particularly of camptothecin production in cultures from *Ophiorrhiza* species. Mitchell Wise outlines how cereal crops can be protected from disease by application of plant defense activators, bioactive compounds that are typically derived from specific plants. Duke et al. provide an extensive review of the history and future prospects of prospecting within the plant kingdom for compounds that protect both crops and humans from insect pests.

The last two perspectives emphasize the role of plant-derived compounds in human health. Zhang et al. review the role of sulfhydryl-reactive compounds, such as the sulforaphanes from broccoli and related plants and several phenolics, including curcumin, in modulating two important pathways that are involved in protecting mammalian cells from oxidative and inflammation-induced damage, such as that which occurs in a number of degenerative diseases and cancer. Eggler and Savinov focus on phytochemicals that are involved in activating the Nrf2 pathway, and thereby help prevent disease.

As always, we hope that you will find these Perspectives to be interesting, informative, and timely. It is our goal that *RAP* will act not only as the voice of the

PSNA, but that it will serve as an authoritative, up-to-date resource that helps to set the gold standard for thought and research in fields related to plant biochemistry.

We welcome suggestions for future articles and comments on the new format.

The RAP Editorial Board

David R. Gang
Laurence B. Davin
Steven C. Halls
Reinhard Jetter
Susan McCormick
J. Fred Stevens
Deyu Xie

Contents

1 Metabolomics Reveals Hemiterpenoids as New Players in the Carbon–Nitrogen Economy 1
Michael H. Beale and Jane L. Ward

2 Electrospray Ionization Traveling Wave Ion Mobility Spectrometry Mass Spectrometry for the Analysis of Plant Phenolics: An Approach for Separation of Regioisomers 21
Fereshteh Zandkarimi, Samanthi Wickramasekara, Jeff Morre, Jan F. Stevens and Claudia S. Maier

3 Camptothecin Production and Biosynthesis in Plant Cell Cultures 43
Takashi Asano, Kazuki Saito and Mami Yamazaki

4 Plant Defense Activators: Application and Prospects in Cereal Crops 55
Mitchell L. Wise

5 Phytochemicals for Pest Management: Current Advances and Future Opportunities 71
Stephen O. Duke, Scott R. Baerson, Charles L. Cantrell, David E. Wedge, Kumudini M. Meepagala, Zhiqiang Pan, Agnes M. Rimando, Kevin K. Schrader, Nurhayat Tabanca, Daniel K. Owens and Franck E. Dayan

6 Sulfhydryl-Reactive Phytochemicals as Dual Activators of Transcription Factors NRF2 and HSF1 95
Ying Zhang, Sharadha Dayalan Naidu, Rumen V. Kostov, Ashley Pheely, Vittorio Calabrese and Albena T. Dinkova-Kostova

7 Chemical and Biological Mechanisms of Phytochemical Activation of NRF2 and Importance in Disease Prevention 121
Aimee L. Eggler and Sergey N. Savinov

Index 157

Contributors

Takashi Asano CREST, Japan Science and Technology Agency, Saitama, Japan

Graduate School of Pharmaceutical Sciences, Chiba University, Chiba, Japan

School of Pharmacy, Iwate Medical University, Iwate, Japan

Scott R. Baerson Natural Products Utilization Research, United States Department of Agriculture, Agricultural Research Service, University, MS, USA

Michael H. Beale The National Centre for Plant and Microbial Metabolomics, Rothamsted Research, Harpenden, Herts, UK

Vittorio Calabrese Department of Chemistry, University of Catania, Catania, Italy

Charles L. Cantrell Natural Products Utilization Research, United States Department of Agriculture, Agricultural Research Service, University, MS, USA

Franck E. Dayan Natural Products Utilization Research, United States Department of Agriculture, Agricultural Research Service, University, MS, USA

Albena T. Dinkova-Kostova Departments of Medicine and Pharmacology and Molecular Sciences, Johns Hopkins University School of Medicine, Baltimore, MD, USA

Division of Cancer Research, Medical Research Institute, University of Dundee, Dundee, Scotland, UK

Stephen O. Duke Natural Products Utilization Research, United States Department of Agriculture, Agricultural Research Service, University, MS, USA

Aimee L. Eggler Department of Chemistry, Villanova University, Villanova, PA, USA

Rumen V. Kostov Division of Cancer Research, Medical Research Institute, University of Dundee, Dundee, Scotland, UK

Claudia S. Maier Environmental Health Sciences Center, Oregon State University, Corvallis, OR, USA

Department of Chemistry, Oregon State University, Corvallis, OR, USA

Kumudini M. Meepagala Natural Products Utilization Research, United States Department of Agriculture, Agricultural Research Service, University, MS, USA

Jeff Morre Department of Chemistry, Oregon State University, Corvallis, OR, USA

Environmental Health Sciences Center, Oregon State University, Corvallis, OR, USA

Sharadha Dayalan Naidu Division of Cancer Research, Medical Research Institute, University of Dundee, Dundee, Scotland, UK

Daniel K. Owens Natural Products Utilization Research, United States Department of Agriculture, Agricultural Research Service, University, MS, USA

Zhiqiang Pan Natural Products Utilization Research, United States Department of Agriculture, Agricultural Research Service, University, MS, USA

Ashley Pheely Division of Cancer Research, Medical Research Institute, University of Dundee, Dundee, Scotland, UK

Agnes M. Rimando Natural Products Utilization Research, United States Department of Agriculture, Agricultural Research Service, University, MS, USA

Kazuki Saito Graduate School of Pharmaceutical Sciences, Chiba University, Chiba, Japan

RIKEN Center for Sustainable Resource Science, Yokohama, Japan

Sergey N. Savinov Purdue University Center for Cancer Research, West Lafayette, IN, USA

Kevin K. Schrader Natural Products Utilization Research, United States Department of Agriculture, Agricultural Research Service, University, MS, USA

Jan F. Stevens Department of Pharmaceutical Sciences, Oregon State University, Corvallis, OR, USA

Environmental Health Sciences Center, Oregon State University, Corvallis, OR, USA

Linus Pauling Institute, Oregon State University, Corvallis, OR, USA

Nurhayat Tabanca Natural Products Utilization Research, United States Department of Agriculture, Agricultural Research Service, University, MS, USA

Jane L. Ward The National Centre for Plant and Microbial Metabolomics, Rothamsted Research, Harpenden, Herts, UK

David E. Wedge Natural Products Utilization Research, United States Department of Agriculture, Agricultural Research Service, University, MS, USA

Samanthi Wickramasekara Department of Chemistry, Oregon State University, Corvallis, OR, USA

Environmental Health Sciences Center, Oregon State University, Corvallis, OR, USA

Mitchell L. Wise Cereal Crops Research Unit, USDA, ARS, Madison, WI, USA

Mami Yamazaki CREST, Japan Science and Technology Agency, Saitama, Japan

Graduate School of Pharmaceutical Sciences, Chiba University, Chiba, Japan

Fereshteh Zandkarimi Department of Chemistry, Oregon State University, Corvallis, OR, USA

Ying Zhang Division of Cancer Research, Medical Research Institute, University of Dundee, Dundee, Scotland, UK

Chapter 1
Metabolomics Reveals Hemiterpenoids as New Players in the Carbon–Nitrogen Economy

Michael H. Beale and Jane L. Ward

Abstract In plant metabolic engineering, a holistic view of plant metabolism can be provided by modern metabolomics. We discuss a recent application of combined nuclear magnetic resonance–mass spectrometry (NMR–MS) metabolomics to hydroponically grown *Arabidopsis thaliana* as a means to study metabolic reprogramming in plant shoots in response to root stress. This led to the discovery of a novel, specifically regulated overflow from the chloroplastic methylerythritol phosphate (MEP) pathway that resulted in the biosynthesis of hemiterpenoid glycosides (HTGs) from hydroxymethylbutenyl diphosphate. The induction of these compounds in leaves is quite specific to a few conditions that lead to low foliar nitrate, implicating the involvement of this anion in the molecular switch. The reprogramming of isoprenoid metabolism in shoots was also correlated with the production of the phenylpropanoids scopolin and coniferin in roots. These results are discussed against a survey of the literature on hemiterpenoids, including isoprene, in order to develop a unified model of metabolic switching of the MEP pathway that allows carbon overflow from a number of points, depending on the species.

1.1 Introduction

In living systems, the metabolites, whether in the free form, or as components of structural and functional macromolecules, are the end products of a complex self-replicating machinery (genes and proteins), which is itself also built from metabolic products. The complement of low-molecular-weight compounds in a living system is known as the metabolome [1]. Metabolomics is the study of the metabolome, usually in context with genetic, physiological, and environmental influences.

In plants, biochemistry is highly geared to assimilate sunlight, carbon dioxide, nitrate, and other nutrients into metabolites and ultimately into biomass. Most metabolic flux is thus dedicated to these primary processes. However, plants are also the source of a rich secondary metabolism, which is highly species dependent and

J. L. Ward (✉) · M. H. Beale
The National Centre for Plant and Microbial Metabolomics, Rothamsted Research, West Common, Harpenden, Herts Al5 2JQ, UK
e-mail: Jane.ward@rothamsted.ac.uk

D. R. Gang (ed.), *50 Years of Phytochemistry Research*,
Recent Advances in Phytochemistry 43, DOI 10.1007/978-3-319-00581-2_1,

has evolved to give plants an ecological advantage in the constant pressure from microbial pathogens and herbivores. One of the key areas of metabolic control is concerned with sensing and adapting to variable carbon (C) and nitrogen (N) supply. Both are crucial for the normal operation of fundamental biochemical and cellular activities, and internal monitoring of C/N balance is critical. It is recognized that C and N metabolism is coordinated. The C/N balance theory [2] was developed to explain the allocation of resources to secondary metabolism for defense. Over the years, there has been much discussion of the veracity of this hypothesis, for example, regarding plant herbivore interactions. Of particular concern has been the inability to predict quantities of individual secondary compounds [3]. However, outside of this discussion on the cost–benefits of plant defense, it is still widely recognized that the primary assimilation of C and N is monitored and balanced [4, 5] and this is supported by transcriptomic studies that demonstrate genome-wide reprogramming of metabolism in response to nitrate levels [6–9]. One of the key applications of plant metabolomics is in the assessment of this type of metabolic reprogramming that occurs when plants are subjected to environmental pressures and/or genetic alterations. An approach to this problem, at the whole plant level (roots and shoots), is described in this perspective.

1.2 Analyzing the Plant Metabolome

The technology used in metabolomics combines data collected using analytical spectroscopies, such as nuclear magnetic resonance (NMR)and mass spectrometry (MS)with that from genomic and/or phenotype studies. Application of multivariate and correlative statistics to these data allows discovery of metabolites, metabolic processes, and genes associated with developmental programming or environmental effects at the tissue or whole organism level. Increasing levels of sophistication in analytical technologies and our ability to generate and utilize data from large sample numbers, as well as in network modeling, have brought metabolomics to the forefront of plant systems biology research [10].

Obviously, metabolism is highly dynamic and the observed metabolome arises from the modulation of the network of biochemical pathways by genetic and environmental influences. Individual metabolite profiles are thus snapshots of the state of the biochemical network. The primary goal of metabolomics analysis is to assay as many metabolites as possible in order to maximize the view of the network. In reality, even when a panel of extraction, separation, and detection strategies is employed, not all metabolites can be detected and a truly holistic measurement of the metabolome cannot be obtained. However, technologies are improving rapidly and well-found laboratories, with access to high-field NMR and a range of MS techniques (especially LC–MS–MS and GC–MS; LC, liquid chromatography; GC, gas chromatography) and ever-growing spectral libraries of authentic metabolites, are able to measure an impressive number of metabolites, especially when the "lipidome"—the large group of homologous compounds for which specific techniques have been developed—is included [11]. It is fair to say that metabolomics data are more meaningful, in terms of tracking the dynamics of the biochemical network in a systems biology approach, when they are collected from time-separated samples of tissue, as the network adjusts to genetic programming or environmental change, including stress, disease, and other predatory pressures.

In plants, there are significant challenges associated with simultaneous analysis of the primary metabolome, which is, qualitatively, fairly consistent across species, and the secondary metabolome which is generally species specific. However, closely related species contain related secondary metabolites (e.g., glucosinolates are associated with the Brassicaceae and isoflavonoids with legumes), and this forms the basis for much of the historical chemotaxonomic literature. The primary metabolome, in green parts of the plant, is intimately linked to photosynthesis and varies diurnally over the day–night cycle. There is also dependence on the growth or developmental stage of the plant. Secondary metabolism can be tissue specific (e.g., trichomes are rich sources of terpenoids) and is often induced by biotic and abiotic stresses. There is also dependence on developmental stage, e.g., older leaves can have different secondary metabolite compositions compared to younger leaves. Thus, plant metabolomics experiments have to be carefully designed to suit the biological question that is to be addressed, and the data interpreted against these background developmental and environmental effects.

In terms of the technologies utilized to measure the plant metabolome, the spectroscopic choices are no different to those faced in human and microbial metabolomics, or indeed in classical natural product discovery and identification. Whether collecting data directly on unpurified solvent extracts, or after separation using chromatography-linked spectroscopic techniques, there are pros and cons for every method in terms of metabolome coverage, dynamic range, accuracy of quantitation, and throughput. A summary of these considerations has been given for *Arabidopsis* [12], but is applicable to most plant systems. There are two different, but complementary, approaches to plant metabolomics. The first involves analysis of tissue extracts against a list of known compounds and, depending on the technique, can give quantitative data on several hundred metabolites, of types ranging from central metabolism to secondary products and lipids. The second approach is to compare metabolite signatures, containing thousands of signals from both known and unknown metabolites, by multivariate statistics to select discriminatory metabolite signals explaining the biology/biochemistry under test. Both approaches are hampered by the lack of standards of known compounds that can be used to annotate metabolite signals (i.e., the MS, NMR, or chromatography peaks). Thus, identification of unknowns and the building of spectral libraries have become major challenges in the quest for maximum coverage of the metabolome. This problem is of course much greater in plants and microbes, where there are large differences in secondary metabolomes.

Here, we describe the use of an *Arabidopsis* experimental setup that allows the rapid assessment of metabolic reprogramming in both roots and shoots when stresses are applied to the root or shoot. The data obtained have brought a root-to-shoot signaling perspective to plant metabolomics, and, thus, a more complete 'systems' view of the reallocation of metabolic resources.

1.3 The *Arabidopsis* Root–Shoot Metabolome Under Stress

A recent publication [13] describing our work on understanding the effects of different stresses on the *Arabidopsis* root and shoot metabolomes highlighted the importance of not confining studies to a single tissue type or time point. A series of experiments utilizing the commercially available Araponics system (http://www.araponics.com) allowed us to develop a reliable protocol for generating *Arabidopsis* root and shoot tissues in enough quantities to allow for a comprehensive metabolomics analysis to be carried out. The system also allowed for easy access of the root system allowing an investigation on the effect of root perturbation on the shoot metabolome and *vice versa*. Early experiments identified common discriminatory metabolites and also revealed key unknown metabolites, the identities of which were determined *via* a classical natural product chemistry approach. The biological provenance of these metabolites was established and results have highlighted a new link with C–N balance and the terpenoid pathway. The work is described in more detail in the subsequent sections of this perspective.

1.3.1 Effect of Total Nutrient Withdrawal

As a relatively simple initial experiment, the effect of total nutrient withdrawal was explored by switching healthy *Arabidopsis* plants, previously grown under full nutrient supply, to water. NMR–MS analysis of extracts of shoot tissue, harvested at two time points (3 days and 7 days), revealed striking and repeatable differences in metabolite fingerprints between nutrient-deficient and nutrient-supplied plants. Increases in common carbohydrates, flavonoids (kaempferol glycosides), and in particular the amino acid phenylalanine and decreases in sinapoyl malate and the amino acids alanine, threonine, aspartate, and glutamate were clearly evident in the ^{1}H NMR data. However, the most striking feature of the NMR data set was the presence of new signals for olefinic hydrogens and aliphatic methyl groups, which were structurally linked to each other and to other signals in the carbohydrate region, in the nutrient-deprived plants. In fact, further inspection of the data indicated that a small family of related compounds had been induced by the nutrient withdrawal, and that there appeared to be associated reprogramming of metabolism involving phenylalanine, flavonoids, and other primary metabolites. Direct infusion electrospray ionization–MS (DI-ESI–MS) data also displayed biomarkers for nutrient deprivation. A decrease in the ion *m/z* 640 (negative ion mode), a novel flavonoid species that was shown to be related to kaempferol 3,7-dirhamnoside (KRR), by MS–MS fragmentation was observed. However, the most prominent feature was in the positive ion fingerprints where two significant ions at *m/z* 287 and 303 were observed only in nutrient-deprived plants.

1.3.2 Identification of the "Unknowns"

MS data collected at higher resolution showed that the ions at *m/z* 287 and 303 actually arose from sodium and potassium adducts of a compound with the empirical formula $C_{11}H_{20}O_7$. Two-dimensional (2D)-NMR experiments on the nutrient-deprived plant extracts indicated that the discriminatory signals identified in the 1D-NMR fingerprints could be assigned to at least two closely related novel glycosides of a C_5-unsaturated diol, present in about a 4:1 ratio. Chemical shift, connectivity, and coupling data indicated that the major compound could be the hemiterpenoid (2*E*)-4-hydroxy-2-methyl-2-buten-1-yl-*O*-D-glucopyranoside (1) and further confirmation of the formula and the presence of two isomers were obtained from GC–MS analysis of the plant material.

The molecular ion was absent but fragments were characteristic of trimethylsilyl glycosides. The key fragment, *m/z* 157, present in both isomers, due to the aglycone had the empirical formula $C_8H_{17}OSi$ and corresponded to the structural fragment $[(CH_3)_3SiOCH_2CH{=}C(Me)CH_2]^-$ that was consistent with the C_5-enediol glycoside structure. 1D nuclear Overhauser enhancement spectroscopy (NOESY) NMR data of the major isomer indicated the *trans (E)* arrangement of the double bond and heteronuclear multiple bond correlation NMR spectroscopy (HMBC) placed the glycosidic linkage at the 1-position. The structures of this compound and the minor isomer were confirmed by synthesis. This was accomplished from (2*E*)-hydroxy-2-methylbut-2-enyl, 4-acetate, an intermediate previously utilized by us in the synthesis of (*E*)-4-hydroxy-3-methyl-but-2-enyl pyrophosphate (HMBPP) [14]. The synthesis involved Koenigs–Knorr coupling of acetylbromoglucose to the 4-monoacetate, followed by deacetylation and, serendipitously, two isomers were produced that were identical to the two hemiterpenoids observed in *Arabidopsis*. The availability of synthetic material enabled comprehensive 2D-NMR investigations of both isomers to be completed in isolation from contaminating plant carbohydrate. This led to the definitive conclusion that the naturally occurring compounds were the two regional isomers of the hemiterpenoid glycoside (HTG)-*(2E)* -2-methylbut-2-en-1,4-diol glycoside, the major isomer being the 1-glycoside (1) and the minor isomer being the 4-glycoside structure (2) as shown in Fig. 1.1. 2-Methylbut-2-en-1,4-diol (3) was a by-product of the synthesis and was also available from separate deacetylation of the starting monoacetate. Small amounts of this compound were also detected in the NMR spectra of nutrient-deprived *Arabidopsis* extracts.

1.3.3 Nitrate, Nitrite, or N? Establishing the Link with HTG Production in Arabidopsis.

To explore the relationship between nutrient deprivation and the formation of HTGs, experiments were carried out where individual nutrients were omitted from the full growth medium. Using metabolomic screening to monitor the production of the identified biomarkers, it was demonstrated that the prime inducer of the HTGs

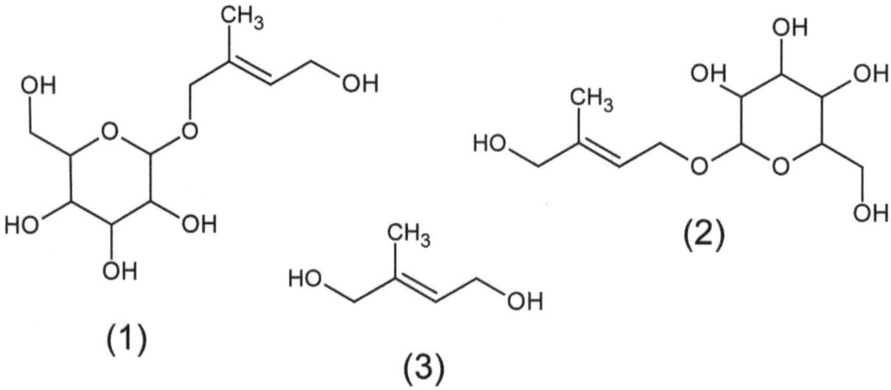

Fig. 1.1 Structures of hemiterpene glycosides identified in *Arabidopsis*

in shoots was root NO_3^- deprivation. The HTGs were still produced when $NH_4^+$was substituted for NO_3^- as a nitrogen source, indicating that nitrate ion sensing was the underlying mechanism behind the synthesis of these compounds. Alternating periods of nitrate deprivation and resupply led to a stepwise accumulation of HTGs under periods of starvation, rising to approximately 1 % dry weight. The HTGs did not appear to be substantially re-assimilated on nitrate resupply. Other primary metabolites also changed during nitrate resupply. Alanine, threonine, aspartate, asparagine, glutamine, and glutamate which were all depleted on nitrate deprivation recovered during the resupply period. Conversely, levels of metabolites that increased during nitrate deprivation (malate, phenylalanine, sucrose, glucose, stachyose, and maltose) were seen to drop during the resupply period, to a level similar to that observed in control plants. Thus, a clear synchronization of certain primary metabolites with nitrate availability and HTG production was observed.

An ion at *m/z* 640, routinely seen in untreated *Arabidopsis* leaf DI-ESI–MS negative ion spectra, was absent in the nitrate-deprived plants. From MS–MS experiments, it was determined that this was formed in the MS source and corresponded to a nitrate adduct of the major flavonoid glycoside KRR and that its intensity was dependent on the nitrate concentration in the leaf tissue. The titer of the *m/z* 640 ion thus serves as a second biomarker for NO_3^- starvation in *Arabidopsis* extracts, and independent measurement of nitrate in the samples by conventional means supported this hypothesis. A weak negative correlation of intensity of ions relating to the HTGs (*m/z* 287 and 303) and *m/z* 640 provided a semi-empirical means of relating high HTG levels directly with low nitrate ion concentration in the same tissue.

It was already apparent that ammonium ion was not involved in the metabolic switch to HTGs. However, there was a possibility that another product of nitrate assimilation—nitrite ion—was the signal. In order to investigate the possibility that levels of nitrite rather than nitrate may drive HTG production, HTG levels in extracts from tissue of nitrate reductase (NR) mutants of *Arabidopsis* were examined. Leaf tissue from these mutants, which cannot assimilate NO_3^-, typically contain higher levels of free nitrate in the foliar tissue. Under starvation conditions the double (*nia1/nia2*) mutant, retaining only 0.5 % of wild-type NR activity [15], did not produce HTGs at all and the levels of nitrate in the foliar tissue remained high presumably as enough nitrate had built up prior to the starvation period (when the plants were growing on nitrate-containing media) which could not then be assimilated. Together, the results pointed to the fact that it was nitrate levels *per se* and not a product of assimilation, such as nitrite or ammonium ion, which were linked with the HTG formation.

1.3.4 Occurrence of the HTGs Under Alternative Stress Conditions and Evidence of Pathway Reprogramming

The Araponics system, and the ability to readily detect the HTG biomarkers, allowed the application of a number of different stresses within the same experiment and measurement of the concentration of HTGs produced. The results of the experiment are shown as a hierarchical cluster analysis in Fig. 1.2. Levels of the HTGs in leaves exceeded 0.5 % dry weight under nitrate deprivation. Wounding of the roots, but not of leaves, induced formation of the HTGs in leaves but other root stresses such as osmotic stress and salt treatment did not induce HTG biosynthesis despite showing characteristic metabolite changes illustrating that these plants were highly stressed. Likewise, cold treatment of the plants did not induce the new metabolite. Potassium deprivation induced HTGs, but the magnitude was less than that of nitrate starvation. However, oxidative stress induced by hydrogen peroxide treatment of roots induced the HTGs in leaves, to levels reaching approximately the same as those observed with nitrate starvation. Most importantly, we determined that high HTG production was always associated with low foliar nitrate levels. In contrast, no HTGs were detected in any of the root samples from this experiment but other major changes in root tissues were apparent. Increases in scopolin and coniferin were not only present in the same samples as those producing HTGs but were also very well correlated with HTG production. These metabolites are products of the phenylpropanoid pathway in *Arabidopsis* and were already known compounds in *Arabidopsis* roots [16].

Although previously associated with light treatment [17], in our study these metabolites were also evidence of a reprogramming of root metabolism that occurs on nitrate deprivation which is in agreement with observations of a scopolin increase in tobacco roots, in response to nitrogen starvation, as early as 1970 [18]. Analogous results on the redirection of carbon flow into phenylpropanoids under nitrate deficiency have also been reported in tobacco leaves [19]. In *Arabidopsis*, data from transcriptomic studies on nitrate limitation are available [6, 9, 20, 21] and these indicate a similar reprogramming, demonstrating that, as chlorophyll biosynthesis was repressed, phenylpropanoid biosynthesis genes were upregulated. Clearly, there is much more work needed to completely understand this relationship. Unpublished work from our laboratory has indicated that HTGs can be induced *via* a

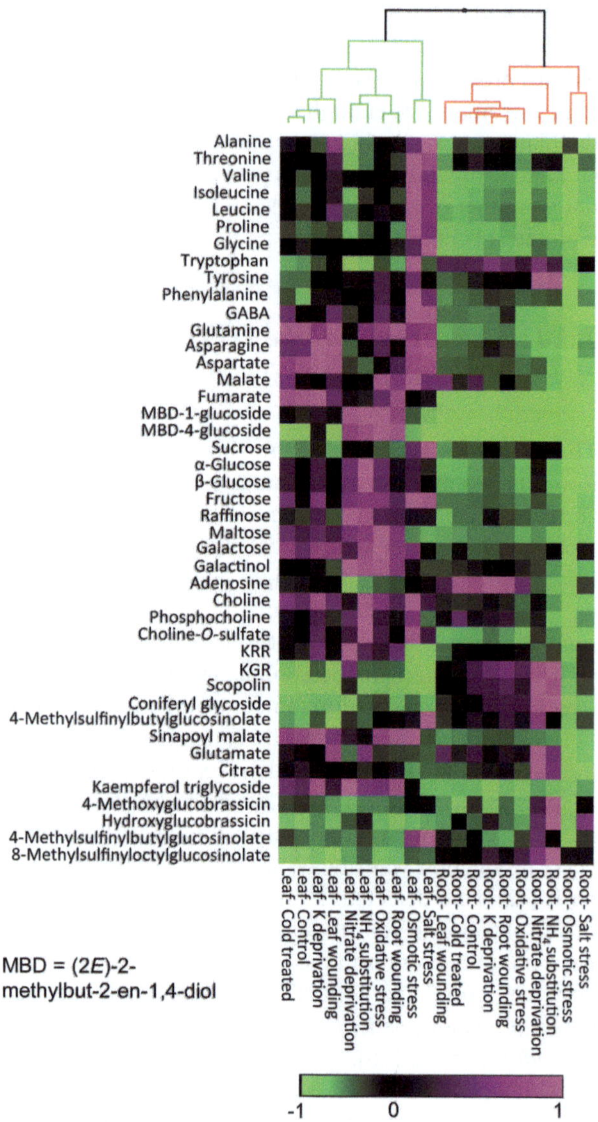

Fig. 1.2 Hierarchical cluster analysis of major metabolites in *Arabidopsis* leaf and root tissue as a result of perturbations from a range of abiotic stresses. Data from ^{1}H NMR (characteristic chemical shift regions) and ESI-MS (*m/z* ions) have been scaled to unit variance and mean centered

high light stress to leaves. Furthermore, despite being supplied with normal nitrate concentrations *via* the roots, foliar nitrate was again very low in the light-stressed plants. Regarding the order of molecular events leading to HTG induction, it is possible that high light induces oxidative stress and as a consequence low foliar nitrate status within the leaves. Alternatively, low nitrate levels caused by an acceleration of photoassimilation in high light induce a pseudo-oxidative stress which in turn triggers the production of HTGs.

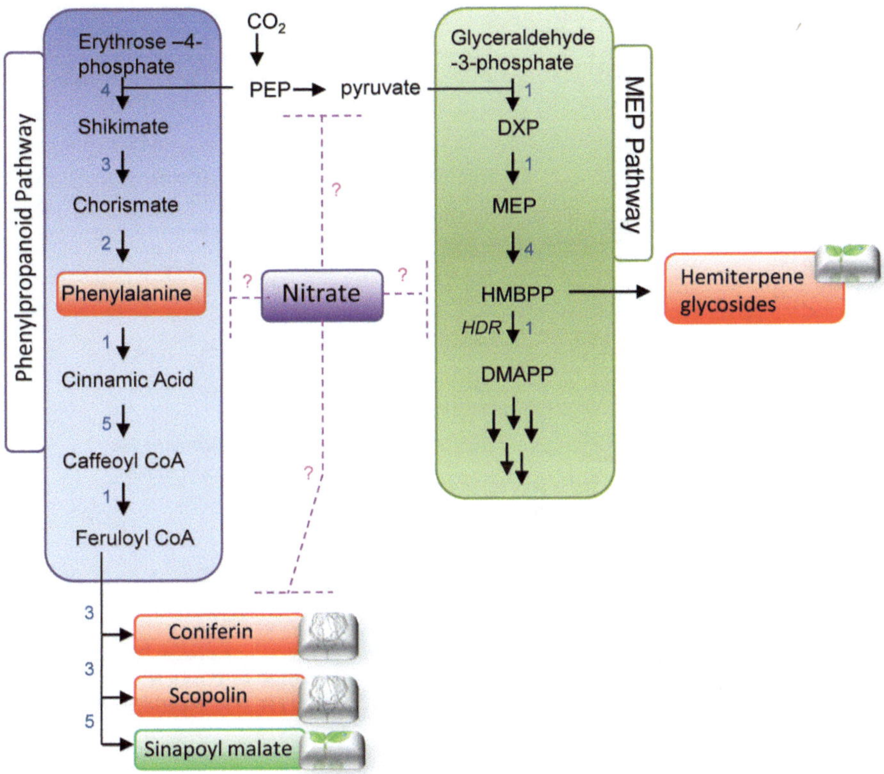

Fig. 1.3 Model of putative pathway interactions in *Arabidopsis*, under foliar nitrate control. Metabolites boxed in red increased under nitrate starvation while those in green decreased. Arrow numbers represent the number of reaction steps between metabolites

The HTGs are products of the chloroplast terpenoid pathway, known as the methylerythritol phosphate (MEP) pathway (see later). The relationship between nitrate and the co-regulation of MEP and phenylpropanoid pathways is shown in Fig. 1.3. The common link between the pathways involves photoassimilates such as phosphoenol pyruvate (PEP) and partitioning of this type of primary precursor may also be part of the nitrate sensing process.

1.4 Hemiterpenoid Diversity

Terpenoids are one of the largest known groups of secondary metabolites. They are incredibly diverse in their structures and biological activities. Despite their structural complexity, however, they all share a common building block, that of the C_5 unit isoprene (4) (Fig. 1.4). While monoterpenes are comprised of two such building

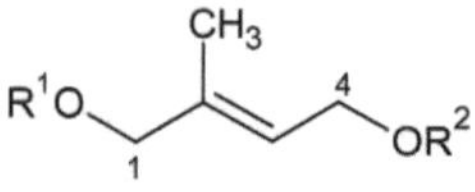

(1) R¹= Glc, R² = H
(2) R¹= H, R² = Glc
(3) R¹= R² = H

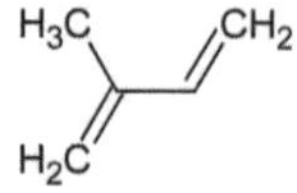

(4) Isoprene

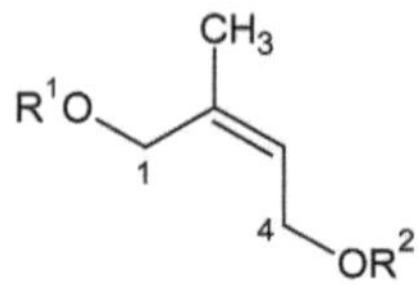

(5) R¹= Glc, R² = H
(6) R¹= H, R² = Glc
(7) R¹= R² = H

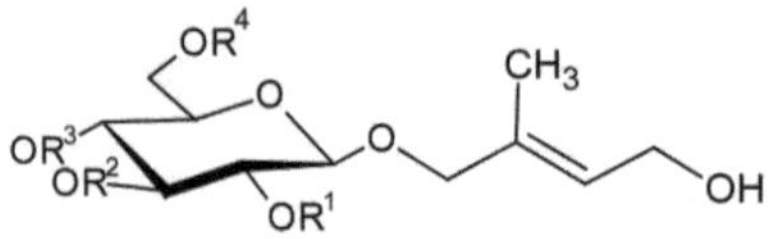

(8) R¹ = R² = 4-HPA, R³ = R⁴ = H
(9) R¹= R² = 3,4-DHPA, R³ = R⁴ = H
(10) R¹= 3,4-DHPA, R² = 4-HPA, R³ = R⁴ =H
(11) R¹= 4-HPA, R² = 3,4-DHPA, R³ = R⁴ = H
(12) R¹= R⁴ = 3,4-DHPA, R² = R³ = H
(13) R¹= R⁴ = 3,4-DHPA, R² = 4-HPA, R³ = H
(14) R¹= 3,4-DHPA, R² = R³ = R⁴ =H
(15) R²= 3,4-DHPA, R¹ = R³ = R⁴ =H
(16) R⁴= 3,4-DHPA, R¹ = R² = R³ =H
(17) R¹ = 4-HPA, R² = R³ = R⁴ = H
(18) R⁴ = 4-HPA, R¹ = R² = R³ = H

HPA = hydroxyphenylacetyl
DHPA = dihydroxyphenylacetyl

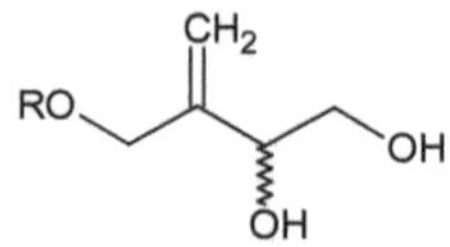

(27) R = glucopyranose
(28) R = allopyranose

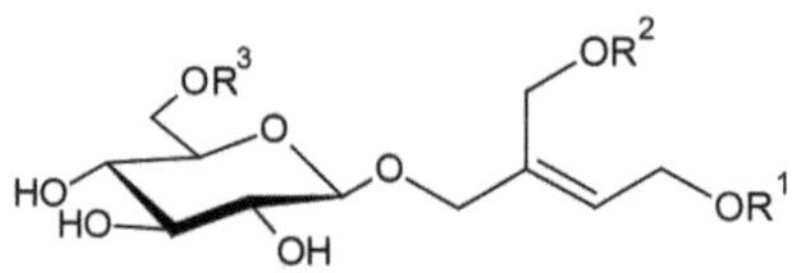

(29) R¹ = caffeoyl, R² = R³ = H
(30) R¹ = R³ = H, R² = caffeoyl
(31) R¹ = R³ = caffeoyl, R² = H

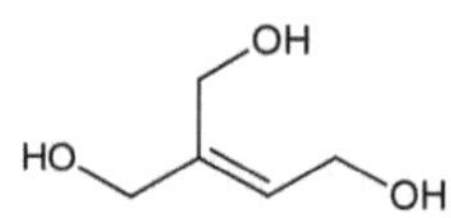

(32)

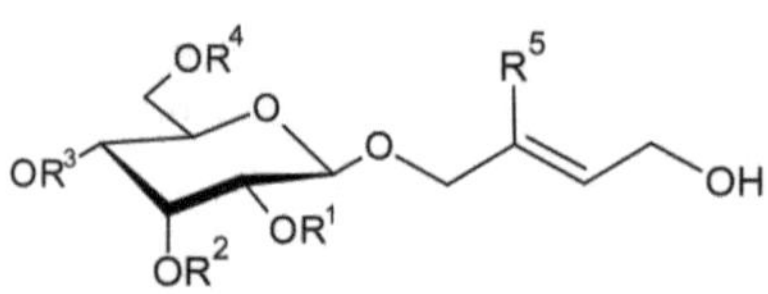

(19) R¹ = R² = R⁴ = H, R³ = caffeoyl, R⁵ = CH_3
(20) R¹ = R² = R³ = H, R⁴ = 3,4-DHPA, R⁵ = CH_2OH
(21) R¹ = R³ = R⁴ = H, R² = caffeoyl, R⁵ = CH_2OH
(22) R¹ = R² = R⁴ = H, R³ = caffeoyl, R⁵ = CH_2OH
(23) R¹ = R² = R³ = H, R⁴ = caffeoyl, R⁵ = CH_2OH
(24) R¹ = R² = R³ = R⁴ = H, R⁵ = CH_2OH
(25) R¹ = R² = R⁴ = 3,4-DHPA, R³ = H, R⁵ = CH_2OH
(26) R¹ = H, R² = R³ = R⁴ = 3,4-DHPA, R⁵ = CH_2OH

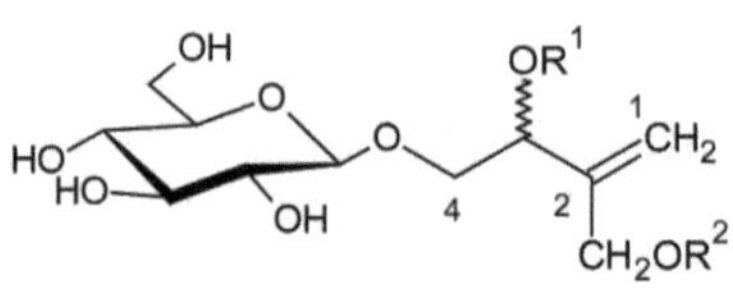

(33) R¹ = H, R² = caffeoyl
(34) R¹ = caffeoyl, R² = H

Fig. 1.4 Structures of plant hemiterpenoids

blocks and sesquiterpenes three isoprene units, hemiterpenes are based on a single C_5 structure. These basic units may of course be saturated or unsaturated, isomerized, or bear functional groups which may themselves allow for conjugation to an array of other structural moieties. The family of hemiterpenoids is quite large but here we discuss key HTG structures which have been previously reported.

The simplest of all hemiterpenes is the compound isoprene (4) which is emitted from the leaves of many plants. How this emission arises and whether it is as a result of a specific function has, over recent years, been the subject of much discussion and speculation [22, 23]. *Arabidopsis*, however, is a non-emitter of isoprene but our results have demonstrated that, under certain conditions, reasonably high levels of the related conjugated hemiterpenoids can be produced in the aerial tissue of the plant. The compounds described earlier had not previously been detected in *Arabidopsis* but have been reported, mostly at low levels, earlier in a number of less-tractable plant species as demonstrated in Fig. 1.4. The 1-*O*-glycoside (1) was initially reported in 1992 [24] and was isolated from the aerial tissue of *Ornithogalum montanum*, a perennial herbaceous bulb found in Italy and the Balkans. The identical compound was reported in the *Torillis japonica* fruit some years later, along with the simple diol (3) in [25]. The 4-*O*-glycoside (2) has also been previously reported as a deterrent of the bean aphid in stems and leaves of *Vicia hirsuta* [26].

Until our work in *Arabidopsis*, the two positional isomers of the glycosylated hemiterpene had not been reported in the same species. This is unlike the corresponding *cis* isomers found in *Vitis vinifera* leaves where both the *cis*-1-*O*-glycoside (5) and the *cis*-4-*O*-glycoside (6) were detected and found to exist in a 3:1 ratio [27]. The *cis*-1-*O*-glycoside (5) was also reported in leaves of the South American cangorosa plant *Maytenus ilicifolia* where it was described as one of three ilicifolinosides isolated [28]. Unlike the *trans* diol (3), the *cis* diol (7) does not appear to have been isolated from plants, but has been synthesized as part of structure determinations of compounds above [26].

Clearly, the common factor with the compounds discussed so far is the aglycone, 2-methyl-but-2-ene-1,4-diol moiety, whether in the *cis* or *trans* configuration. Further, more complex analogs based on the same agylcone skeleton have also been reported from methanolic extracts of the Japanese Fern, *Hymenophyllum barbatum* [29]. These analogs (8–18) include a *trans* 1-*O*-substituted glucose moiety. The family of compounds isolated, named hymenosides, included a mixture of mono- and dihydroxyphenylacetylated glucose substituents with the position of the substitution varying around the carbohydrate ring. A much wider selection of these so-called hymenosides, bearing a hydroxylated aglycone skeleton and a substituted allose moiety, were also isolated from the same plant species [29–31]. These compounds bore a mixture of caffeoyl- or dihydroxyphenylacetyl-substituted allose arrangements and were conjugated to either 2-methylbut-2-ene-1,4-diol (19), 2-hydroxymethyl-but-2-ene-1,4-diol (20–26) or 2-methylene-butane-1,3,4-triol (27–28). Caffeoyl-substituted hemiterpene glycosides have also been isolated from the bark and roots of two different *Ilex* (holly) species within the Aquifoliaceae family. Three new caffeoyl-substituted HTGs were isolated from methanolic extracts of the dried bark of *Ilex macropoda* [32]. These metabolites (29–31), based on the triol structure (32), bear caffeoyl substituents at one (C-4 or C-5) or two (C-4 and C-5) positions alongside the glucose substitution at C-1. Two further HTGs, named pubescenosides, were isolated from the roots of *Ilex pubescens* [33]. These compounds (33, 34) possessed potent antiplatelet aggregation activities and were based on the aglycone structure 2-hydroxymethyl-3,4-dihydroxy-but-1-ene.

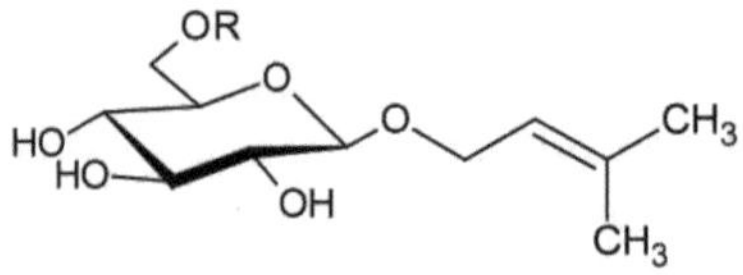

(35) R = H
(39) R = apiofuranose

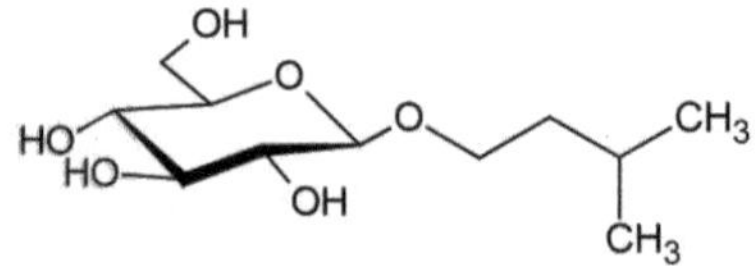

(36)

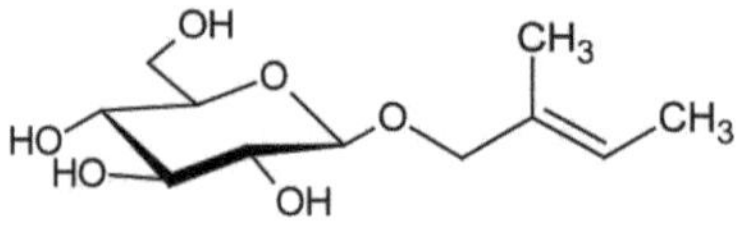

(37)

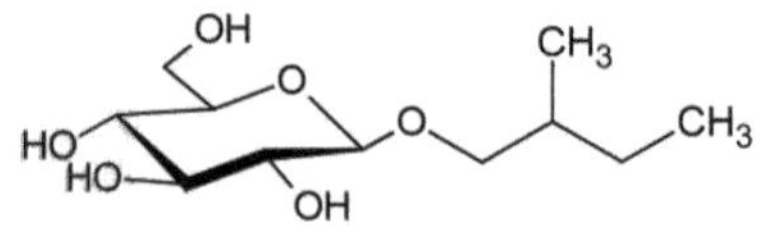

(38)

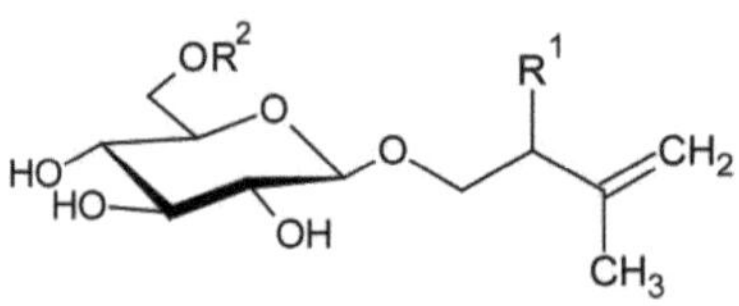

(40) R^1 = H, R^2 = apiofuranose
(42) R^1 = OH, R^2 = H

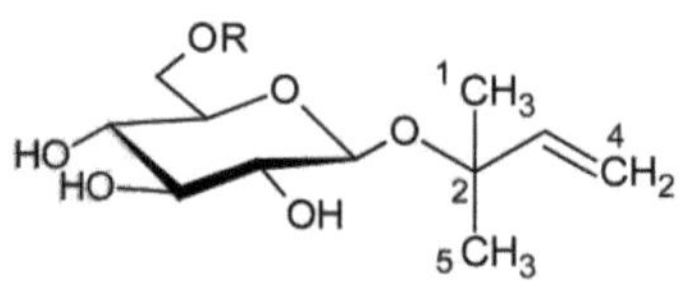

(41) R = apiofuranose

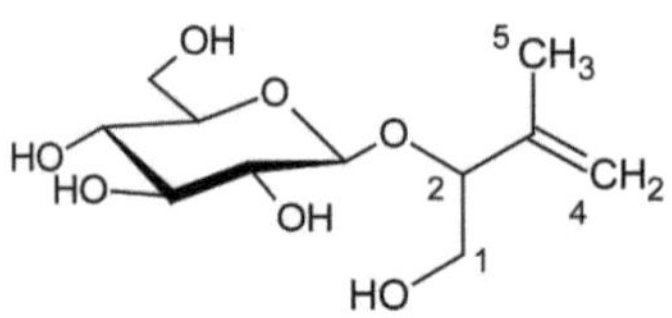

(43)

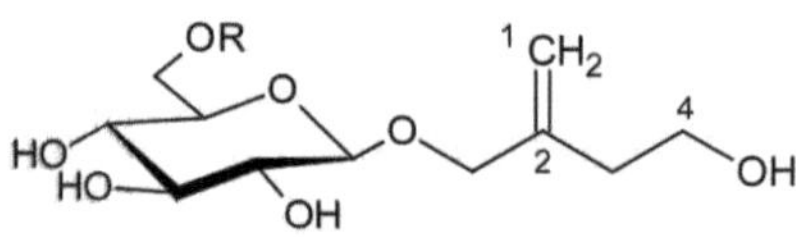

(44) R = H
(45) R = CO-C(CH_2)CH_2CH_2OH

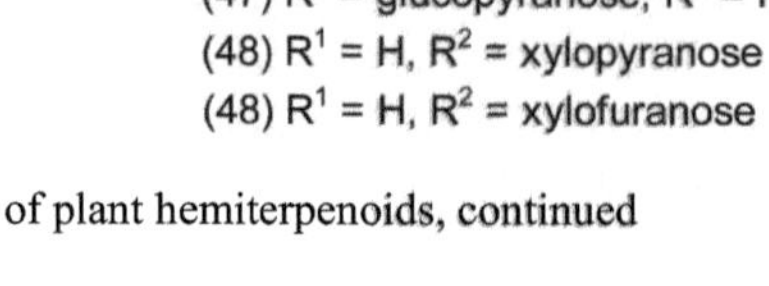

(46) $R^1 = R^2$ = H
(47) R^1 = glucopyranose, R^2 = H
(48) R^1 = H, R^2 = xylopyranose
(48) R^1 = H, R^2 = xylofuranose

Fig. 1.5 Structures of plant hemiterpenoids, continued

Further HTGs (Fig. 1.5) have been reported in a number of diverse species. Perhaps the most relevant of these is a prenyl glycoside based on a 3-methylbut-2-en-1-ol (dimethylallyl alcohol) (35). Originally reported as a fragrance precursor in flower buds of *Citrus unshiu* [34] and subsequently in 1977 [35] *via* Soxhlet extraction of the aerial parts of *Ferula loscosii,* a prevalent species in Spain, this metabolite was again reported as being part of a collection of metabolites present in the water-soluble fraction of the methanol extract of fennel fruit (*Foeniculum vulgare,* Umbelliferae) alongside the fully saturated analog (36) and the positional isomer, 2-methyl-but-2-en-1-ol glucoside (37), and the analog (38) [36]. Early reports of these saturated HTGs generally associated these types of compounds with a variety of medicinal properties such as antimicrobial and antifungal activities. Such an example [37] was the isolation from leaves of the Canarian shrub *Bystropogon plumosus* of 2-methylbutan-1-yl-β-D-glucoside (38) to levels of around 0.01 %.

An apiofuranose analog (39) of the dimethylallylalcohol glycoside (35) was isolated from *Vitis vinifera* wine [38] alongside the similar analog from the 3-methylbut-3-en-1-ol glycoside (40). An apiofuranose analog (41) of 2-methylbut-3-en-2-ol glycoside has also been isolated from the water-soluble extract of the root and rhizome tissue of *Glehnia littoralis* [39]. The position of the double bond within the aglycone structure varies and there have been a number of reports of such metabolites with a terminal double bond structure. The position of the double bond and subsequent substitution are presumably as a result of the biosynthetic route utilized in particular tissues. Aerial tissue of *Lamium album* afforded the 2-*O*-glucoside of 3-methylbut-3-en-1,2-diol (43) named by the authors as hemialboside [40] while the extraction of fresh rhizomes of *Coptis japonica* var. *dissecta* allowed the isolation of a group of close analogs (42, 44–45) [41]. Finally, a series of hemiterpene glycosides (46–49) based on the 3-methylbut-3-en-1-ol parent structure were isolated recently from fruits of noni (Indian mulberry) [42]. This set of metabolites included the 1-*O*-glucoside (46) and conjugates further extended with an additional glucopyranose (47), xylopyranose (48), or xylofuranose (49).

In summary, we have highlighted examples of compounds that appear to be structurally related to the main C_5 building blocks of isoprenoid biosynthesis. Other related structures have been summarized by Dembitsky [43]. Of particular interest in metabolomics and carbon flow is the provenance of the variety of skeleta observed, and this is discussed in the next section.

1.5 Hemiterpenes—Products of Internal Regulation of the Chloroplast Isoprenoid Pathway?

The structural resemblance of the hemiterpene glycosides to HMBPP, a key intermediate in the chloroplastic MEP pathway, indicates a biosynthetic pathway involving hydrolysis of the diphosphate group followed by glycosylation at either end of the 1,4-diol. (Fig. 1.6). Support for this route was provided [13] by feeding intermediates—HMBPP, desoxyxylulose phosphate (DXP), and 1,4-diol—to leaves excised from nitrate-depleted *Arabidopsis* plants. Although, in feeding studies such as these, phosphorylated compounds are unlikely to be taken up into plant chloroplasts, the results indicated that the distal precursor DXP was taken up and

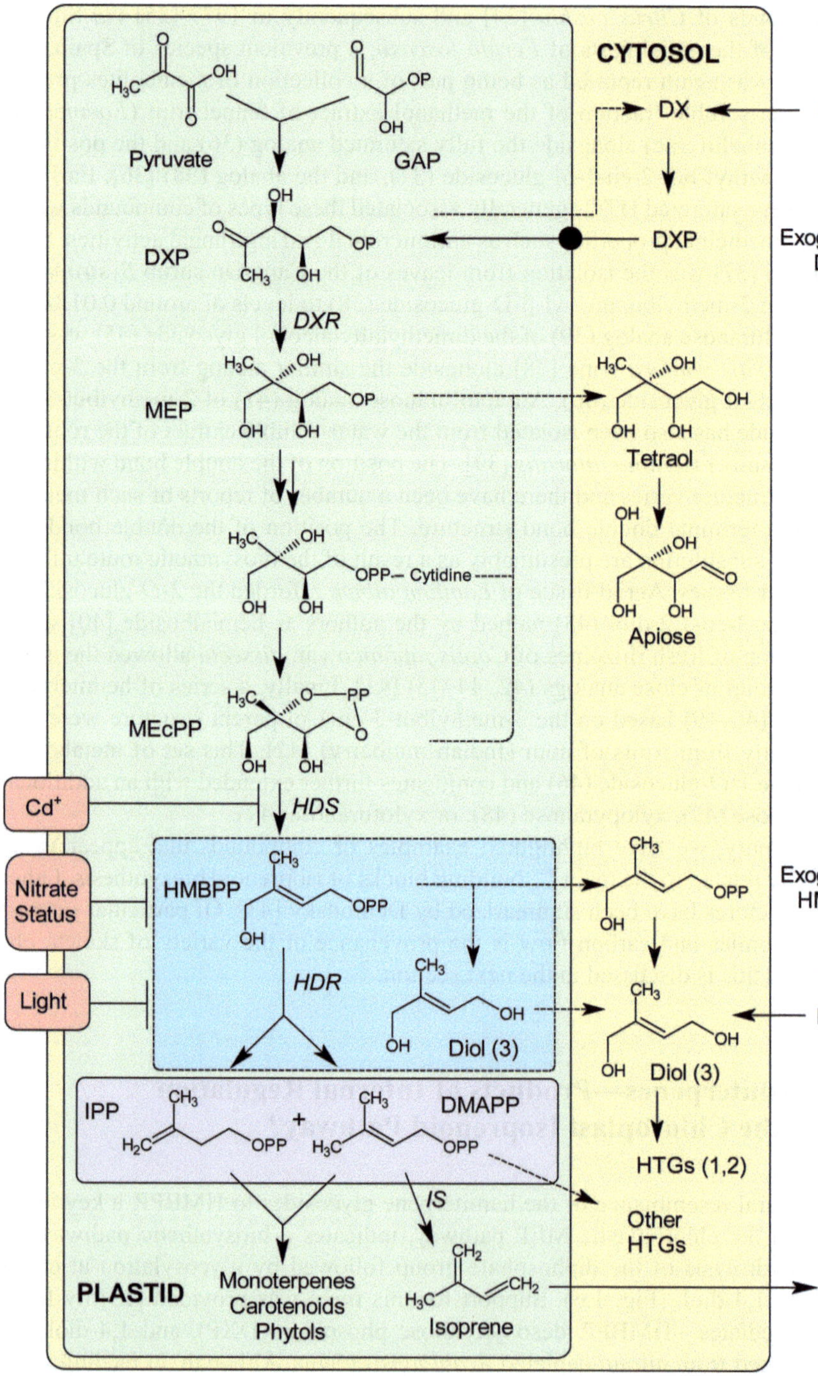

Fig. 1.6 The chloroplast MEP pathway and proposed provenance of hemiterpene alcohols and derivatives *via* stress-induced pathway shunts

resulted in the formation of some 3–4 times more hemiterpene glycosides in nitrate-depleted tissue when compared with controls. There are precedents for DXP uptake and turnover in plant tissue and the mechanism is thought to involve extracellular dephosphorylation to the alcohol and then uptake into the cell, followed by re-phosphorylation to DXP which is then actively transported into the chloroplast by a carrier protein [44, 45]. When either HMBPP or the 1,4-diol was fed to cut leaves of nitrate-depleted *Arabidopsis*, Ward et al. [13] observed high turnover of both compounds to the hemiterpene glycosides, providing firm evidence for the proposed pathway. However, there was no difference in turnover between nitrate-depleted and control plants.

Given that HMBPP has been shown not to be transported across chloroplast membranes [46] and it can be assumed that it also was dephosphorylated outside of the cell, this result indicated that the glycosylation of the diol occurred outside of the chloroplast. Together, the feeding results from nitrate-depleted plants support the hypothesis that nitrate signaling decreased flux through the enzyme hydroxymethylbutenyl diphosphate reductase (HDR; Fig. 1.6) causing HMBPP to build up and overflow to the HTGs. HMBPP generated *de novo* within the chloroplast (from the endogenous pathway or from exogenous DXP) was subject to the nitrate-regulated shunt mechanism. HMBPP added exogenously does not appear to reach the chloroplast but is dephosphorylated and glycosylated irrespective of nitrate levels. The experiments could not distinguish between low-nitrate induction of an HMBPP phosphatase in the chloroplast and the alternate mechanism of low-nitrate-induced translocation of excess HMBPP from chloroplast to cytosol (Fig. 1.6). The results observed in *Arabidopsis* parallel those observed in virus-induced gene silencing of HDR reductase activity in *Nicotiana benthamiana* [47]. In this system, silencing of HDR led to the conversion of HMBPP to the diol, via the monophosphate HMBP, indicating similar regulatory mechanisms to those observed in *Arabidopsis*.

Many of the known hemiterpene structures depicted in Figs. 1.4 and 1.5 can be accommodated by similar biosynthetic conversions from HMBPP, dimethylallyl diphosphate (DMAPP), or isopentenyl diphosphate (IPP). Some structures (e.g., 41, Fig. 1.5) can arise from allylic rearrangement of the double bond in the dephosphorylation step or at the alcohol stage. Most difficult to rationalize is the biosynthesis of the dihydro analogs (e.g., 36, Fig. 1.5) and *cis*-diol-based structures (e.g., 5–7, Fig 1.4). Both appear to arise out of the downstream modification of dimethylallyl alcohol by either hydrogenation or hydroxylation.

Other naturally occurring C_5 polyols such as apiose and the related tetraol (Fig. 1.6) are also potentially formed from MEP pathway intermediates as shown [25]. This opens up the possibility of other overflow mechanisms from different places in the MEP pathway. It has also been reported that 2-C-methyl-D-erythritol 2,4 cyclodiphosphate, a potential precursor of such tetraol structures, builds up in high light, high temperatures, and also on cadmium stress in spinach tissue [48]. The production of isoprene itself is well established from chloroplastic DMAPP *via* the action of a specific enzyme isoprene synthase and has been suggested to be a shunt product when DMAPP pool size becomes too large [22]. There has been much discussion concerning the raison d'être for isoprene emission from plants. This occurs in many tree species and contributes to greenhouse gases as well as the summer haze over forests. There is evidence that isoprene emission has a function in the protection of plant membranes under heat and light stress [23], but recent results of field trials with poplar trees where isoprene synthase genes have been suppressed by RNA interference (RNAi) techniques indicate that there is no growth and development advantage from isoprene emission, and perhaps the only benefit of isoprene emission may come from interactions with herbivores [49]. Despite the different routes of formation of isoprene (induction of a specific chloroplast enzyme) and the HTGs (potentially via nonspecific extra-chloroplastic phosphorylases and glycosyl transferases), there are parallels that can be drawn. Both processes appear to be governed by the relevant isoprenoid diphosphate pool size [13, 50] and furthermore appear to be a consequence of plant nitrate status [13, 51] as well as induced by high light [52, Ward et al. unpubl.].

Obviously more work needs to be done on flux through the MEP pathway, but this work and increasing evidence in the literature indicate that hemiterpenoid production is the result of a multilevel intermediate pool size management system that is responsive to a number of external and internal signals.

1.6 Future Prospects

This work has demonstrated the power of unbiased metabolomics not only to discover new natural products but also to provide a "systems" overview of metabolism as it adjusts to pressure from environmental and nutritional stress. Until this work, flux through the terpenoid pathway in *Arabidopsis* was thought to be low, relative to many plants that accumulate large amounts of secondary terpenoids. Despite the presence of greater than 30 mono- and sesquiterpene synthase genes [53], oxidized terpenes have not been reported in the model plant, although volatile mono- and sesquiterpene hydrocarbons are emitted at low levels by flowers and roots [54]. Aside from the extensive research on isoprene, hemiterpenoids have been a somewhat neglected research area. However, the discovery of HTGs in the model plant and the specific nature of the induction of their biosynthesis will stimulate renewed interest in these molecules. The possibility that the HTG formation in *Arabidopsis* is a paradigm for isoprene production in trees will bring a greater range of genetic tools and resources to bear on carbon overflow mechanisms. This will be of direct interest to terpene engineering whether the strategic outlet is the accumulation of valuable compounds or for manipulation of interactions with predators. The production of isoprene and other linear terpenoids as biofuels is also a growing research topic. Genetic engineering is beginning to shed light on the 'raison d'être' for isoprene production by tree species [49], and the debate on carbon overflow versus thermoprotectant continues. Delineation of the regulatory mechanism of HTG production in *Arabidopsis* will surely inform on this discussion.

The use of the metabolomics screen for HTGs will allow a detailed genetic, temporal, and environmental study of the induction of the low foliar nitrate state that leads to HTG synthesis. Nitrate sensing and signaling in leaf biochemistry has long been mooted as a key factor in metabolic control in plants [7]. The work described here presents a new opportunity to explore this, using the extensive genetic resources available for *Arabidopsis*. The links between high light, oxidative stress, and low foliar nitrate revealed in this study can also now be explored by using HTG production as readout. The high correlation between HTG production in the shoots with diversion of phenylpropanoid flow to coniferin and scopolin in the roots perhaps indicates that aspects of the plant defense response cross over with the nitrate response, and that nutrient limitation may activate the flow of carbon into secondary defense metabolites.

In the discussion, we have suggested that a palette of regulatory mechanisms modulate the chloroplastic MEP pathway. We suggest that different plants use different routes to off-load excess flux in stress conditions. The determinant appears to be the pool size of the relevant isoprenoid diphosphate. *Arabidopsis* and tobacco [47] produce HMBPP-derived products. Isoprene is emitted by many tree species, and DMAPP pool size appears to be a factor [50] in isoprene emission. Similarly, hemiterpenoids apparently derived from DMAPP (and IPP) have been widely reported (Figs. 1.4 and 1.5). Also of interest are the tetraols that could be derived from a number of the higher precursors (Fig. 1.6). Future developments in metabolomic analysis of the MEP pathway metabolites will be necessary as an aid to understanding the extent of this form of regulation. Also of interest, in the biosynthesis of HTGs, are the dephosphorylation and glycosylation steps. Evidence for the presence or absence of specific and inducible enzymes for these steps needs to be sought and combined transcriptomic–metabolomic studies of stressed plants will aid in this endeavor.

Acknowledgments Rothamsted Research receives grant-aided support from the Biotechnology and Biological Sciences Research Council of the UK.

References

1. Oliver SG, Winson MK, Kell DB, Banganz F (1998) Systematic functional analysis of the yeast genome. Trends in Biotech 16:373
2. Bryant JP, Chapin FS, Klein DR (1983) Carbon/nutrient balance of boreal plants in relation to vertebrate herbivory. Oikos 40:357
3. Hamilton JG, Zangerl AR, DeLuca EH, Berenbaum MR (2001) The carbon-nutrient balance hypothesis: its rise and fall. Ecol Lett 4:86
4. Stitt M (1999) Nitrate regulation of metabolism and growth. Curr Opin Plant Biol 2:178
5. Zheng ZL (2009) Carbon and nitrogen nutrient balance signalling in plants. Plant Signal Behav 4:584–91
6. Scheible WR, Morcuende R, Czechowski T, Fritz C, Osuna D, Palacios-Rojas N, Schindelsach D, Thimm O, Udvardi MK, Stitt (2004) Genome-wide reprogramming of primary and secondary metabolism, protein synthesis, cellular growth processes, and the regulatory infrastructure of Arabidopsis in response to nitrogen. Plant Physiol 136: 2483

7. Vidal EA, Gutierrez RA (2008) A systems view of nitrogen nutrient and metabolite responses in Arabidopsis. Curr Opin Plant Biol 11:521
8. Coruzzi GM, Zhou L (2001) Carbon and nitrogen sensing and signaling in plants: emerging 'matrix effects' Curr Opin Plant Biol 4:247
9. Wang R, Okamoto M, Xing X, Crawford NM (2003) Microarray analysis of the nitrate response in Arabidopsis roots and shoots reveals over 1,000 rapidly responding genes and new linkages to glucose, trehalose-6-phosphate, iron and sulfate metabolism. Plant physiol. 132:556
10. Redestig H, Szymanski J, Hirai MY, Selberg J, Willmitzer L, Nikoloski Z, Saito K (2011) Data integration, metabolic networks and systems biology. In: Hall RD (ed) Annual Plant Reviews, vol43, Biology of Plant Metabolomics. Wiley-Blackwell, Oxford, pp 261–316
11. Welti R, Shah J, Li W, Li M, Chen J, Burke JJ, Fauconnier M-L, Chapman K, Chye M-L, Wang X (2007) Plant lipidomics: Discerning biological function by profiling plant complex lipids using mass spectrometry. Frontiers in Bioscience 12:2494
12. Beale MH, Sussman MR (2011) Metabolomics of *Arabidopsis Thaliana*. In: Hall RD (ed) Annual Plant Reviews, vol43, Biology of Plant Metabolomics. Wiley-Blackwell, Oxford, pp 157–180
13. Ward JL, Baker JM, Llewellyn AM, Hawkins ND, Beale MH (2011) Metabolomic analysis of *Arabidopsis* reveals hemiterpenoid glycosides as products of a nitrate ion-regulated, carbon flux overflow. Proc Nat Acad Sci 108:10762
14. Ward JL, Beale MH (2002) Synthesis of (*2E*)-4-hydroxy-3-methylbut-2-enyl diphosphate, a key intermediate in the biosynthesis of isoprenoids. J Chem Soc Perkin Trans I 6:710
15. Wilkinson JQ, Crawford HM (1993) Identification and characterisation of a chlorate-resistant mutant of *Arabidopsis thaliana* with mutations in both nitrate reductase structural genes *NIA1* and *NIA2*. Mol Gen Genet 239:289
16. Bednarek P, Schneider B, Svatosˇ A, Oldham NJ, Hahlbrock K (2005) Structural complexity, differential response to infection, and tissue specificity of indolic and phenylpropanoidsecondary metabolism in Arabidopsis roots. Plant Physiol138:1058
17. Hemm MR, Rider SD, Ogas J, Murry DJ, Chapple C (2004) Light induces phenylpropanoid metabolism in Arabidopsis roots. Plant J 38:765
18. Armstrong GM, Rohrbaug LM, Rice EL, Wender SH (1970) Effect of nitrogen deficiency on concentration of caffeoylquinic acids and scopolin in tobacco. Phytochemistry 9:945
19. Fritz C, Palacios-Rojas N, Feil R, Stitt M (2006) Regulation of secondary metabolism by the carbon-nitrogen status in tobacco: NO3- inhibits large sectors of phenylpropanoid metabolism. The Plant Journal 46:533
20. Bi YM, Wang RL, Zhu T, Rothstein SJ (2007) Global transcription profiling reveals differential responses to chronic nitrogen stress and putative nitrogen regulatory components in Arabidopsis. BMC Genomics 8:Article Number: 281
21. Peng MS, Bi YM, Zhu T, Rothstein SJ (2007) Genome-wide analysis of Arabidopsis responsive transcriptome to nitrogen limitation and its regulation by the ubiquitin ligase gene NLA. Plant Mol Biol 65:775
22. Sanadze GA (2004) Biogenic isoprene (a review). Russ J Plant Physiol 51:729
23. Sharkey TD, Wiberley AE, Donohue AR (2008) Isoprene emission from plants: Why and how. Ann Bot (Lond) 101:5
24. Nicoletti M, Tomassini L, Foddai S (1992) A new hemiterpene glucoside from *Ornithogalum montanum*. Planta Med. 58:472
25. Kitajima J, Suzuki N, Ishikawa T, Tanaka Y (1998a) New hemiterpenoid pentol and monoterpenoid glycoside of *Torillis japonica* fruit and consideration of the origin of apiose. Chem Pharm Bull 46:1583
26. Ohta N, Mori N, Kuwahara Y, Nishida R (2006) A hemiterpene glucoside as a probing deterrent of the bean aphid, *Megoura crassicauda*, from a non-host vetch, *Vicia hirsute*. Phytochemistry 67:584
27. Messerer M, Winterhalter P (1995) (*2Z*)-4-Hydroxy-2-methyl-2buten-1-yl-β-D-glucoside and (*2Z*)-1-Hydroxy-2-methyl-2buten-4-yl-β-D-glucoside: Two new hemiterpene glucosides from *Vitis vinifera* leaves. Nat Prod Lett 5:241

28. Zhu N, Sharapin N, Zhang J (1998) Three glucosides from *Maytenus Ilicifolia*. Phytochemistry 47:265
29. Toyota M, Oiso Y, Asakawa Y (2002) New glycosides from the Japanese fern *Hymenophyllum barbatum*. Chem Pharm Bull 50:508
30. Toyota M, Oiso Y, Asakawa Y (2001) New bitter-tasting hemiterpene glycosides from the Japanese Fern *Hymenophyllum barbatum*. Chem Pharm Bull 49:1567
31. Oiso Y, Toyota M, Asakawa Y (2001) Hymenosides A-F, six new hemiterpene glucosides from the Japanese fern *Hymenophyllum barbatum*. Chem Pharm Bull 49:126
32. Fuchino H, Tachibana H, Tanaka N (1997) Three new hemiterpene glycosides from *Ilex macropoda* Chem Pharm Bull 45:1533
33. Jiang Z-H, Wang J-R, Li M, Liu Z-Q, Chau K-Y, Zhao C, Liu L (2005) Hemiterpene glucosides with anti-platelet aggregation activities from *Ilex pubescens* Chem Pharm Bull 68:397
34. Yoshikawa K, Kobayashi M, Arihara S (1996) Flower fragrance precursors from flower buds of *Citrus unshiu* Marcov Natural Med 50:176
35. Pinar M, Martin-Lomas M (1977) 2-Methyl-3-buten-2-yl-β-D-glucopyranoside from *Ferula loscosii*. Phytochemistry 16:281
36. Kitajima J, Ishikawa T, Tanaka Y (1998b) Water-soluble constituents of Fennel I. Alkyl glycosides. Chem Pharm Bull 46:1643
37. Nahrstedt A, Economou D, Wray V (1990) 2-Methylbutan-1-yl-β-D-glucoside. A hemiterpene glucoside from *Bystropogon plumosus*. J Nat Prod 53:1387
38. Baltenweck-Guyot R, Trendel J-M, Albrecht P (1997) New hemiterpene glycosides in *Vitis vinifera* wine. J Nat Prod 60:1326
39. Kitajima J, Okamura C, Ishikawa T, Tanaka Y (1998c) New glycosides and furocoumarin from the *Glehnia littoralis* root and rhizome. Chem Pharm Bull 46:1939
40. Damtoft S, Jensen SR (1995) Hemialboside, a hemiterpene glucoside from *Lamium album*. Phytochemistry 39:923
41. Yoshikawa K, Kinoshita H, Arihara S (1997) Non-basic components of *Coptis rhizome* II. Four new hemiterpenoid glucosides. Two new phenylpropanoid glucosides and a new flavonoid glycoside from *Coptis japonica* var. *dissecta*. Nat Med (Tokyo) 51:244
42. Akihisa T, Seino K-I, Kaneko E, Watanabe K, Tochizawa S, Fukatsu M, Banno N, Metori K, Kimura Y (2010) Melanogenesis inhibitory activities of iridoid-, hemiterpene-, and fatty acid glycosides from the fruits of *Morinda citrifolia* (Noni). J Oleo Sci 59:49
43. Dembitsky VM (2006) Astonishing diversity of natural surfactants: 7. Biologically active hemi- and monoterpenoid glycosides. Lipids 41:1
44. Schwender J, Zeidler J, Gröner R, Müller C, Focke M, Braun S, Lichtenthaler FW, Lichtenthaler HK. (1997) Incorporation of 1-deoxy-d-xylulose into isoprene and phytol by higher plants and algae. FEBS Lett 414:129
45. Hemmerlin A, Tritsch D, Hartmann M, Pacaud K, Hoeffler J-F, van Dorsselaer A, Rohmer M, Bach TJ. (2006) A cytosolic Arabidopsis D-xylulose kinase catalyzes the phosphorylation of 1-Deoxy-D-Xylulose into a precursor of the plastidial isoprenoid pathway. Plant Physiol 142:441
46. Flügge UI, Gao W (2005) Transport of isoprenoid intermediates across chloroplast envelope membranes. Plant Biol 7:91
47. Page JE, Hause G, Raschke M, Gao W, Schmidt J, Zenk MH, Kutchan TM (2004) Functional analysis of the final steps of the 1-deoxy-D-xylulose 5-phosphate (DXP) pathway to isoprenoids in plants using virus-induced gene silencing. Plant Physiol 134:1401
48. Rivasseau C, Seeman M, Boisson AM, Streb P, Gout E, Douce, R, Rohmer, M, Bligny R (2009) Accumulation of 2-C-methyl-d-erythritol 2,4-cyclodiphosphate in illuminated plant leaves at supra optimal temperatures reveals a bottleneck of the prokaryotic methylerythritol 4-phosphate pathway of isoprenoid biosynthesis. Plant Cell Environ 32:82
49. Behnke K, Grote R, Brüggemann N, Zimmer I, Zhou G, Elobeid M, Janz D, Polle A, Schnitzler J-P. (2012) Isoprene emission-free poplars—a chance to reduce the impact from poplar plantations on the atmosphere. New Phytologist 194:70

50. Vickers CE, Possell M, Hewitt CN, Mullineaux PM (2010) Genetic structure and regulation of isoprene synthase in Poplar (Populussp). Plant Mol Biol 73: 547
51. Rosensteil TN, Ebbets AL, Khatri WC, Fall R, Monson RK (2004) Induction of poplar leaf NO_3 reductase: a test of extra chloroplastic control of isoprene emission. Plant Biol 6:12
52. Botella-Pavia P, Besumbes O, Phillips MA, Carretero-Paulet L, Boronat A, Rodriguez-Concepcion M (2004) Regulation of carotenoid biosynthesis in plants: evidence for a key role of hydroxymethylbutenyl diphosphate reductase in controlling the supply of plastidial isoprenoid precursors. Plant J 40:188
53. Aubourg S, Lecharny A, Bohlmann J (2002) Genomic analysis of the terpenoid synthase (AtTPS) gene family of *Arabidopsis thaliana.* Mol Gen Genomics 267:730
54. D'Auria JC, Gershenzon J (2005) The secondary metabolism of *Arabidopsis thaliana*: growing like a weed. Curr Opin Plant Biol. 8:308–16

Chapter 2
Electrospray Ionization Traveling Wave Ion Mobility Spectrometry Mass Spectrometry for the Analysis of Plant Phenolics: An Approach for Separation of Regioisomers

Fereshteh Zandkarimi, Samanthi Wickramasekara, Jeff Morre, Jan F. Stevens and Claudia S. Maier

Abstract The use of ion-mobility spectrometry (IMS) coupled to mass spectrometry (IMS–MS) for biomolecule analyses has steadily increased over the past two decades, and is now applied to both proteomic and metabolomic investigations. This chapter describes the application of traveling-wave ion-mobility spectrometry–mass spectrometry (TWIMS–MS) to the analysis of a selection of bioactive phytochemicals used in dietary supplements. Applications include the analysis of grape seed proanthocyanidins and the structural characterization of bioactive constituents of dietary supplements using TWIMS-MS in conjunction with tandem mass spectrometry. We also discussed is the application of TWIMS-MS for the gas-phase mobility separation of structural isomers and the estimation of collision cross sections for a small selection of phenolic compounds from hop. Recent applications of IMS–MS to a broad range of biomolecule measurements have demonstrated that IMS–MS has emerged as a powerful analytical technique capable of providing the separation space necessary to analyze highly complex samples. We give a perspective on emerging applications of IMS–MS for small molecule and biopolymer applications. The combination of devices that allow real-time monitoring of living systems using IMS–MS is an exciting avenue of facilitating system-biology experiments. The future of IMS–MS is bright and full of opportunities.

C. S. Maier (✉) · F. Zandkarimi · S. Wickramasekara · J. Morre
Department of Chemistry, Oregon State University, Corvallis, OR 97331, USA
e-mail: Claudia.maier@oregonstate.edu

S. Wickramasekara · J. Morre · J. F. Stevens · C. S. Maier
Environmental Health Sciences Center, Oregon State University, Corvallis, OR 97331, USA

J. F. Stevens
Department of Pharmaceutical Sciences, Oregon State University, Corvallis, OR 97331, USA

Linus Pauling Institute, Oregon State University, Corvallis, OR 97331, USA

D. R. Gang (ed.), *50 Years of Phytochemistry Research*,
Recent Advances in Phytochemistry 43, DOI 10.1007/978-3-319-00581-2_2,

2.1 Introduction

There is increasing evidence that plant phenolics have health benefits which may, at least partially, stem from their antioxidant and radical scavenging activity [1–3]. Considering the increasing interest in plant phenolics as nutraceuticals, comprehensive profiling methods for plant extracts are highly needed. We report on the characterization of plant phenolics using traveling-wave ion-mobility spectrometry–mass spectrometry (TWIMS–MS) and emphasize the structural analysis of plant secondary metabolites that are commonly found in over-the-counter dietary supplements.

The use of ion-mobility spectrometry (IMS) coupled to mass spectrometry (IMS–MS) for biomolecule analyses has steadily increased since the 1990s. Many applications describe IMS–MS for studying peptides and proteins and their folding behaviors in the gas phase [4–10]. IMS–MS has also been used for assessing synthetic polymers [11, 12]. More recently, IMS–MS has been described as a powerful addition to the arsenal of tools for the structural analysis of small molecules including drugs, metabolites, lipids, carbohydrates, phytochemicals, and other natural products [13–19]. Comprehensive reviews are available that describe in detail the principles and applications of IMS–MS [20–22]. Briefly, in IMS ions are separated according to their charge state, shape, and size. IMS systems function as gas-phase separation devices. IMS uses nondestructive low-energy collisions to separate ions predominately on the basis of ion-neutral collision cross sections. Ion-mobility separations in the gas phase have considerably lower resolution compared to the resolution that can be achieved with modern condensed phase chromatographic separation technologies. However, the separation of ions occurs several orders of magnitudes faster than separations based on liquid chromatographic techniques; ion-mobility separations usually occur on a time scale of milliseconds compared to the seconds to hours in chromatographic separations. MS measurements occur in the microsecond range and, as such, are nested within the IMS experiments [23]. IMS–MS experiments allow real-time separations of the components of complex mixtures and provide access to three-dimensional (3D) analytical information, namely shape, mass, and abundance. The combination of IMS with MS results in two-dimensional plots of drift time (t_d) versus *m/z*. The three-dimensionality of TWIMS–MS datasets is best captured in so-called driftscope images that contain information regarding the drift time (t_d, in ms) and *m/z* values displayed in a nested fashion, t_d as function of (*m/z*), with ion abundances given in a color-coded style. These images enable the extraction of distinct features that otherwise would get lost or overlap in crowed spaces of traditional mass spectra. A unique feature of IMS–MS is the ability to conduct drift time measurements that allow the calculation of collision cross sections (CCSs) for low- and high–molecular-weight molecules and the possible separation of isomeric analytes, which is not possible solely with MS-based techniques [13–15].

IMS–MS systems are available in many different instrument configurations. In the traditional configuration, the drift tube-based IMS device is placed between the ionization source and the mass analyzer [20]. The recent advent of a commercial

IMS–MS system, in which a traveling-wave IMS device has been integrated into a hybrid quadrupole-orthogonal acceleration time-of-flight (TOF) mass spectrometer, opens new possibilities for the structural characterizations of small molecules and biopolymers [24]. In this contribution, we describe the application of TWIMS–MS for the analysis of a selection of bioactive phytochemicals used in dietary supplements. We first give a brief description of the TWIMS–MS instrument used. Then, we discuss several applications of TWIMS–MS: (1) the analysis of a biopolymer mixture, namely grape seed proanthocyanidins; (2) the structural characterization of bioactive constituents of dietary supplements using TWIMS in conjunction with tandem mass spectrometry; and (3) the gas-phase mobility separation of structural isomers and the estimation of collision cross sections for a small selection of phenolic compounds from hop. We conclude with a brief review of recent developments and applications, and provide a perspective on the emerging application of IMS–MS for small molecule and biopolymer applications.

2.2 Ion-Mobility Mass Spectrometry Using an Electrospray Ionization Quadrupole Traveling-Wave Ion-Mobility TOF Instrument

Many different instrument designs have been described that combine IMS with different types of MS analyzers [20]. The availability of a commercial hybrid system that integrates traveling-wave ion-mobility separation with a quadrupole TOF analyzer makes the technology accessible to a wider research community. For the applications described here, we used a Waters Synapt G2 HDMS instrument equipped with an electrospray ionization (ESI) source. This instrument has the following configuration: a quadrupole mass filter, the TriWave™ section consisting of three traveling-wave (T-wave) devices (Trap T-wave, Ion mobility separation (IMS) T-Wave, and Transfer T-wave) and an orthogonal acceleration (oa) TOF analyzer (Fig. 2.1). The traveling-wave IMS device is a radio-frequency (RF) ion guide based on a stack of ring electrodes. The RF voltage confines the ions radially. By applying a continuous series of DC voltage pulses traveling along the stacked ring electrodes, ions are moved through the gas-filled guide. The ability of an ion to travel along with the DC pulse through the gas-filled T-wave ion guide will depend on the ion's size, charge, shape, and the interaction cross section between the ion and background gas. Instrumental parameters that affect the ability of an ion to move with the traveling wave are the amplitude of the wave, the travel velocity of the wave, and the type and pressure of the background gas. Applying an optimized set of parameters, high-mobility ions will travel with the wave and pass through the ion guide faster than ions with low mobility, which roll over the top of the wave and as a result spend more time in the device [24].

The TriWave™ section of this instrument consists of three traveling-wave-enabled stacked ring ion guides. This configuration allows for unique fragmentation experiments (Table 2.1). The trap ion guide is used for the accumulation of ions and

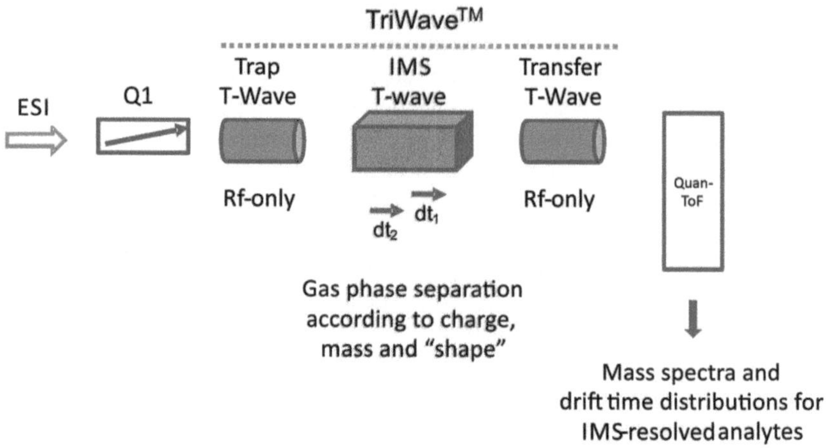

Fig. 2.1 Conceptual diagram of the commercially available TWIMS–MS instrument (Waters Synapt G2 HDMS instrument) operated in the mobility time-of-flight (TOF) mode. This operating mode is used for gas-phase mobility separation of ions in combination with high-resolution mass spectrometry. This mode also enables the extraction of drift times for the estimation of collision cross sections after calibrating the traveling-wave ion-mobility separator. In mobility-TOF mode, the quadrupole (Q1) analyzer is operated in the transmitting mode, the trap and transfer devices serve at radio-frequency (Rf)-only ion guides. *T-wave*, traveling wave; *IMS*, ion mobility spectrometry; *dt*, drift time

release of ions as packets into the ion-mobility separation device. The transfer ion guide conveys the mobility-separated ions to the orthogonal acceleration (oa) TOF analyzer. Fragmentation can take place either in the trap, in the transfer device, or in both devices. A detailed description of the working principle and the design of the traveling-wave ion-mobility separator have been published previously by Giles et al. [24, 25]. Details on the theoretical background on classical IMS–MS and the adaption to traveling TWIMS–MS have also been described in several recent papers [24–27]. In the following sections, we describe the use of TWIMS–MS experiments for the structural characterization of plant phenolics.

2.3 TWIMS–MS Analysis of Biopolymers: Application to Grape Seed Proanthocyanidins

Grape seed extracts have been extensively studied by diverse mass spectrometric techniques. In most cases, the mass spectrometric analyses were accompanied by laborious and extensive chromatography of the highly complex grape seed proanthocyanidin mixtures [28–30]. Considering the current interest in grape seed proanthocyanidins as nutraceuticals in biomedical applications, comprehensive and fast profiling of grape seed extracts is highly desirable. The analysis of proanthocyanidins by ESI–MS is challenging due to the overlapping of ion signals of constituents

Table 2.1 Operating modes of the TWIMS–MS instrument utilizing the TriWave™ section

Mode	TriWave™ usage			Products
	Trap	TWIMS device	Transfer	
Mobility-ToF	Ion guide only	Ion-mobility separation	Ion guide only	Ions are separated according to their mobility; drift time measurements enable collision cross-section estimations after calibration of the TWIMS device
Trap fragmentation	Elevated energy	First-generation fragment ions	Ion guide only	First-generation fragment ions are ion mobility separated
Transfer fragmentation	Storage device only	Precursor ions	Elevated energy	Precursor ions are separated according to their mobility. First-generation fragment ions align with precursor ion drift time
Time-aligned parallel (TAP) fragmentation	Elevated energy	First-generation fragment ions	Elevated energy	First-generation fragment ions are separated according to their mobilities. Activation in the transfer device results in second-generation product ions which are time-aligned to the respective first-generation product ion precursor

of the highly complex mixtures of proanthocyanidins oligo- and polymers differing in length (i.e., degree of polymerization or DP), subunit composition, and type of linkage between flavanol units [29–33].

The three-dimensionality of TWIMS–MS dataset is highlighted in Fig. 2.2a–d. ESI–TWIMS–MS driftscope images (drift time versus m/z; color coding: white most to blue least abundant ions) of a typical grape seed proanthocyanidin preparation are shown (Figs. 2.2a, c). What makes these images remarkable is (i) the separation of the ions into distinct charge state groups which are denoted as (+1) and (+2) and (ii) the separation of the proanthocyanidin oligomers into distinct ion clusters. The sodium adducts of the proanthocyanidin oligomer ions displayed shorter drift times than the corresponding protonated molecular ions (Fig. 2.2a).

In the TWIMS–MS driftscope image depicted in Fig. 2.2a, the gas-phase mobility separation of singly protonated proanthocyanidin ions from the doubly protonated proanthocyanidin ions is highlighted. The two charge state groups are denoted with (+1) and (+2). In Fig. 2.2b, a section of the ESI mass spectrum (*m/z* 500–1,400) is depicted. Singly charged ion signals dominate the mass spectrum. Due to the mobility separation of singly and doubly charged ions, it is possible to extract distinct ion clusters, which helps with the analysis of overlapping ion signals. For instance, extraction of the singly protonated ions of the procyanidin trimer (PC3, $[M+H]^+$, m/z 867.2) and the doubly protonated ions of the procyanidin hexamer (PC6, $[M+2H]^{2+}$, m/z 866.2) results in mass spectra that show baseline resolved isotope clusters (Fig. 2.2c).

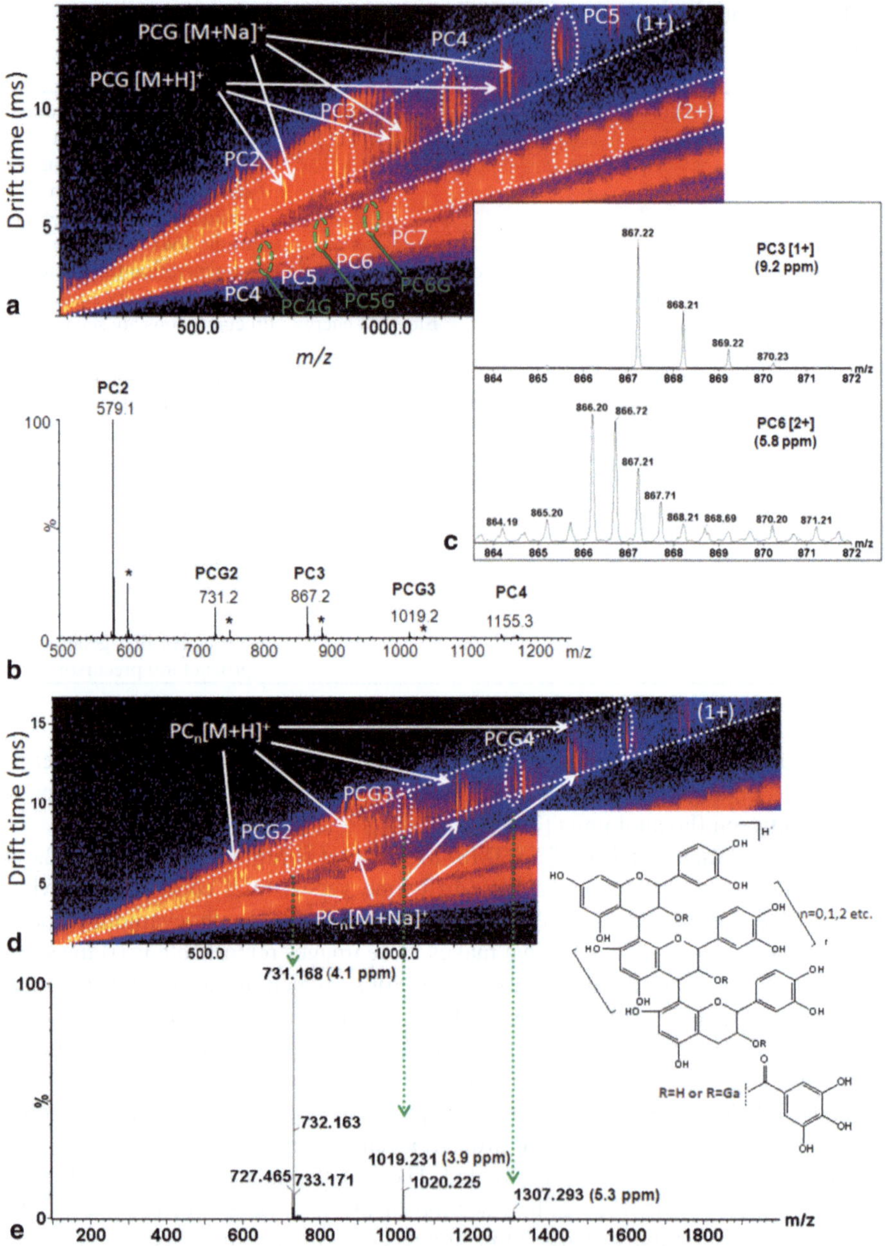

Fig. 2.2 TWIMS–MS analysis of grape seed proanthocyanidins. **a** TWIMS–MS driftscope image (drift time versus *m/z*) of a grape seed extract. PC ions separate in the TWIMS cell into charge groups labeled with (+1) and (+2) representing singly protonated and doubly protonated ions. **b** ESI mass spectrum of grape seed proanthocyanidins (depicted is range from *m/z* 500–1,450). The *asterisk* indicates the sodium adduct, $[M+Na]^+$, of the respective protonated molecular ion,

In the image depicted in Fig. 2.2d, arrows mark the singly protonated molecular ion clusters of procyanidins at *m/z* 579, 867, 1,155, and 1,443 corresponding to procyanidins with increasing degree of polymerization, namely DP2, DP3, DP4, and DP5. The sodiated molecular ions $[M+Na]^+$ are marked as well. The procyanidin monogallate ion clusters are encircled in this plot. Extraction of selected ion clusters allows the detailed analysis of ion signals that belong to a distinct proanthocyanidin series. For instance, the ion clusters of the procyanidin monogallates (PCG) were extracted and the respective mass spectrum of the extracted ion clusters is shown in Fig. 2.2b. The spectrum shows only the ion signals of procyanidin monogallates with DP2 (MH^+, *m/z* 731.17), DP3 (MH^+, *m/z* 1,019.23), and DP4 (MH^+, *m/z* 1,307.29).

Although the present TWIMS–MS spectra were obtained in the positive ionization mode, proanthocyanidins give information-rich TWIMS–MS plots in the negative mode. Matrix-assisted laser desorption ionization (MALDI)-MS analysis has been described as a powerful approach for the characterization of PC mixtures [28, 30]. Therefore, it would be interesting to see if the combination of MALDI with TWIMS–MS would further advance the analysis of these highly complex biopolymer mixtures. The above example attempts to demonstrate some of the features that TWIMS–MS offers as an analytical platform for the in-depth interrogation of proanthocyanidin preparations. We anticipate that TWIMS–MS will emerge as measurement technology for the comprehensive analysis of other biopolymers and bio-inspired plastics as well.

Experimental Details The instrument was operated in positive ionization mode with an ESI capillary voltage of 2.75 kV and a sampling cone voltage of 30 V. The other conditions were as follows: extraction cone voltage, 4.0 V; ion source temperature, 100 °C; desolvation temperature, 300 °C; desolvation gas flow rate, 500 L/h; and cone gas flow rate, 40.0 L/h. Ion-mobility separation conditions included: ion-mobility gas flow rate, 75.30 mL/min, IMS wave velocity, 542 m/s, and wave height, 40.0 V. Argon was used as collision gas in the trap and transfer cells, while nitrogen (N_2) was used as IMS cell gas. Data acquisition was carried out using Waters MassLynx (V4.1), and for IMS data processing DriftScope software (V 2.1, Waters) was used. Positive ion mass spectra were acquired in the resolution mode over a mass range of 100–2,500 *m/z* using continuum mode setting. Mass calibration in positive mode was performed by infusing sodium iodide solution (2 µg/µL, 1:1 (v/v) water:2-propanol).

$[M+H]^+$. **c** The inset shows the extracted mass spectra of the singly protonated ions of procyanidin trimers, PC3, (upper mass spectrum) and the doubly protonated ions of procyanidin hexamers, PC6 (lower mass spectrum). Note the baseline-resolved isotope cluster for the doubly protonated PC6 ions; TWIMS–MS enables the gas-phase separation of the singly protonated ions from the doubly protonated PC oligomers avoiding overlapping of the isotope clusters on the *m/z* scale. **d** Distinct ion clusters can be individually extracted and exported to display the respective mass spectra. The extracted mass spectrum depicting singly protonated procyanidin (PC_n) oligomers with DP 2–4 is shown. Sodiated proanthocyanidin ion $[M+Na]^+$ has been annotated as well. Procyanidin monogallate (PCG) ion clusters are seen in between procyanidin clusters. **e** Extracted ion signals for procyanidin monogallates

2.4 Tandem Mass Spectrometry Approaches for the Structural Analysis of Plant Phenolics Using Dried Spot Analysis in Combination with TWIMS–MS

The need for high-throughput techniques for the analysis of dietary supplements and active ingredients encouraged us to explore the combination of dried spot analysis using thin-layer chromatography (TLC) plates in combination with ESI–TWIMS–MS. A combination of TLC, desorption electrospray ionization, and TWIMS–MS has been described previously for the direct analysis of pharmaceutical formulations [34]. The combination of ion-mobility separation with MS allows gas-phase ions to be separated by their mobility and then to be analyzed according to their mass-to-charge ratio in the TOF analyzer. Analysis specificity is further increased by combining ion-mobility separation with collision-induced fragmentation in the transfer region of the Synapt G2 instrument, thus, enabling the extraction of structural information and high resolution accurate mass measurements in one experiment. The combination of TLC-based spot analyses and ESI–TWIMS–MS resembles a multidimensional separation approach that results in high-content mass spectral information for the analytes of interest. Here, we describe the application of TWIMS–MS with tandem mass spectrometry for the analysis of bioactive flavonoids in dietary supplements, namely rutin (quercetin-3-O-rutinoside, $C_{27}H_{30}O_{16}$, M_{mono} 610.1534 Da) and hesperidin (hesperitin-7-O-rutinoside, $C_{28}H_{34}O_{15}$, M_{mono} 610.1898 Da).

Dietary supplements were extracted with methanol and the extracts were spotted on cellulose TLC plates. A CAMAG TLC MS interface combined with an LC system was used for eluting the analytes from the TLC plate and subsequent infusion into the mass spectrometer. Under the experimental conditions used, this approach enabled the interrogation of dietary supplement spots for a time period of approximately 1.5 min. A typical total ion response is shown in Fig. 2.3a. By extracting all ion signals at the plateau of the total ion chromatogram, a 3D driftscope image is generated (Fig. 2.3b). Note the relatively broad ion distribution at *m/z* 611 indicating insufficient resolution to separate the protonated ions of hesperidin and rutin under the TWIMS conditions used. However, two well-separated ion signal distributions were observed for the sodiated ions, $[M+Na]^+$, of those two flavonoid glycosides (*m/z* 633). The selected ion signals with *m/z* 633 show better signal-to-noise ratios than the protonated molecules $[M+H]^+$ at *m/z* 611 (Fig. 2.3d, e). Under the TWIMS conditions used, hesperidin ($[M+Na]^+$, *m/z* 633.18) and rutin

---

measured under the current conditions (*m/z* range 100–1,200 Da). Note the broad ion assemblies at *m/z* 611 and the two ion distributions with similar *m/z* values at *m/z* 633 but clearly different drift time distributions. **d** and **e** Comparison of drift time distributions of ion signals observed for two over-the-counter dietary supplements that contain both flavonoid diglycosides, hesperidin ([M+Na] *m/z* 633.18, dt 5.21 ms), and rutin ([M+Na] *m/z* 633.14, dt 5.97 ms). Drift time distributions were obtained by selecting the ions at *m/z* 633 with the quadrupole device Q1 followed by gas-phase separation of the ions in the TWIMS cell. In **e**, the mass spectrum has a different *m/z* scale than in **d** to illustrate that the ion signals for both flavonoid diglycosides ($[M+Na]^+$) are observable at the *m/z* scale

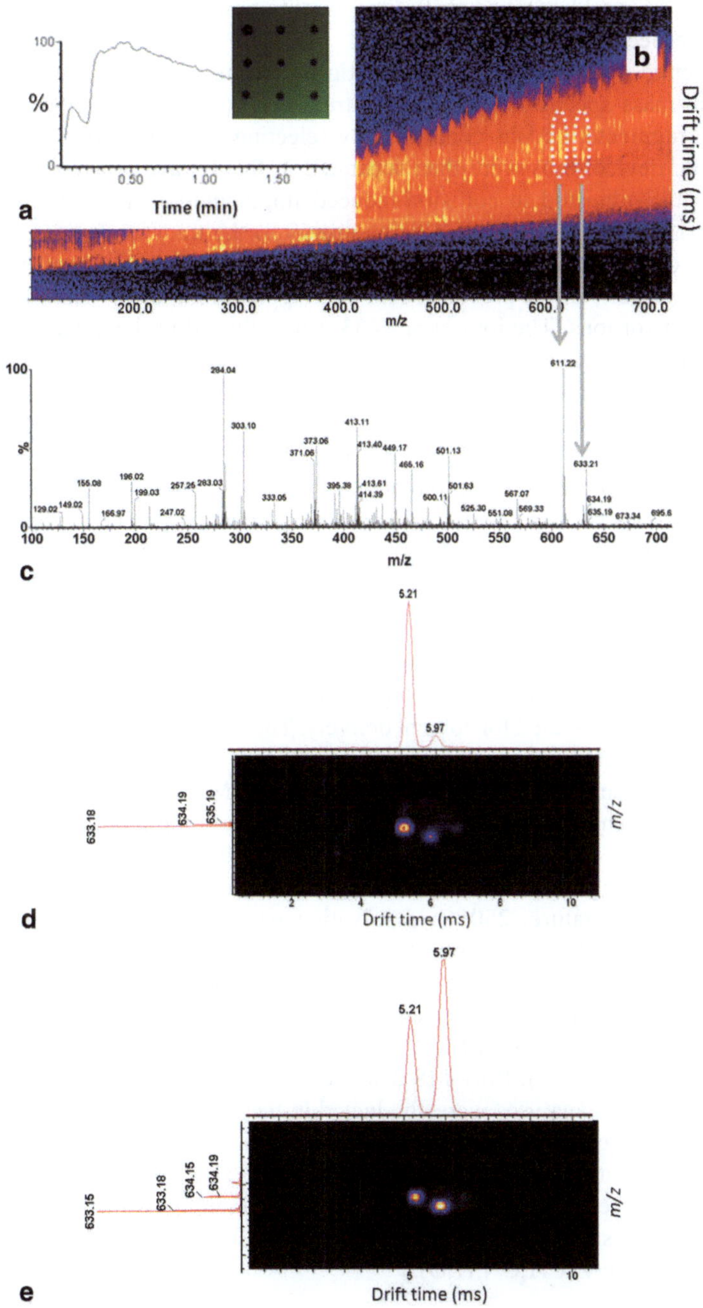

Fig. 2.3 Dried spot analysis of phytochemicals in dietary supplements with electrospray ionization TWIMS–MS. **a** Total ion chromatogram of a dietary supplement spotted on cellulose TLC plate, extracted from the plate, and infused into the mass spectrometer. **b** Driftscope image (drift time versus *m/z*) and **c** the integrated mass spectrum over the entire range of drift time distributions

([M+Na]$^+$, *m/z* 633.14) had drift time distributions centered around 5.21 and 5.97 ms, respectively.

In order to obtain structural information, the ions at *m/z* 633 were subjected to tandem mass spectrometry using the transfer region of the TriWave™ device. Transfer fragmentation was conducted by selecting the ions at *m/z* 633 using the quadrupole device. Ions were then separated in the T-wave ion-mobility cell and subsequently subjected to collision-induced fragmentation in the transfer region (Fig. 2.4a). Applying elevated collision energy to the transfer device causes fragmentation of the mobility-separated precursor ions. Because the fragment ions preserve their velocity of the precursor ion, the fragment ions align with the drift times of the precursor ions. The ions at *m/z* 633 were selected in the quadrupole region, separated in the T-wave cell, and subsequently fragmented by collisions in the transfer region. The integrated fragment ion spectrum is shown in Fig. 2.4b. In Figs. 2.5 and 2.6, the fragment ion mass spectra of the sodiated molecular ions of hesperidin and rutin are shown, respectively. The time-aligned and compound-specific ions were extracted for each of the flavonoid glycosides separately. Because transfer dissociation experiments were conducted using sodiated precursor ions, the fragment ions are sodiated as well. Fragment ions of the rutinoside moiety dominated the spectrum for both species. The observed fragment ions are indicated in the schematic presentation of the two flavonoid glycoside structures.

Experimental Details The analytes were extracted using a CAMAG TLC interface from cellulose TLC plates using 75 % acetonitrile/25 % water. A Shimadzu LC-10AD pump was used for solvent delivery. The flow rate was 0.1 mL/min. Mass spectral experiments were performed using a Waters Synapt G2 HDMS mass spectrometer (Manchester, UK) equipped with TWIMS. Mass spectra were acquired in positive mode. The instrument was operated in the resolution mode with a capillary voltage of 3.0 kV and a sampling cone voltage of 30.0 V. The other conditions comprise the following: extraction cone voltage, 4.1 V; ion source temperature, 80 °C; desolvation temperature, 250 °C; desolvation gas flow rate, 500 L/h; and cone gas flow rate, 5.0 L/h. Ion-mobility separation conditions included ion-mobility gas flow rate; 90 mL/min, wave velocity ramping from 480 to 556 m/s; and wave height, 40.0 V. Argon was used as collision gas on the trap and transfer cells, while nitrogen (N_2) was used as IMS cell gas. Data acquisitions were carried out using Waters MassLynx (v4.1). IMS–MS data were processed with DriftScope software (v2.1, Waters). All analyses were conducted in the positive ionization mode. Mass spectra were acquired over the mass range of 50–1,200 *m/z* in continuum mode. A 0.1 ng/μL solution of leucine enkephaline ([M+H]+556.2771) was infused at 5 μL/min as the reference mass (lock mass) for accurate mass measurements. Mass calibration in positive mode was performed by infusing sodium formate (5 mM, prepared in 1:1 (v/v) CH_3CN:H_2O).

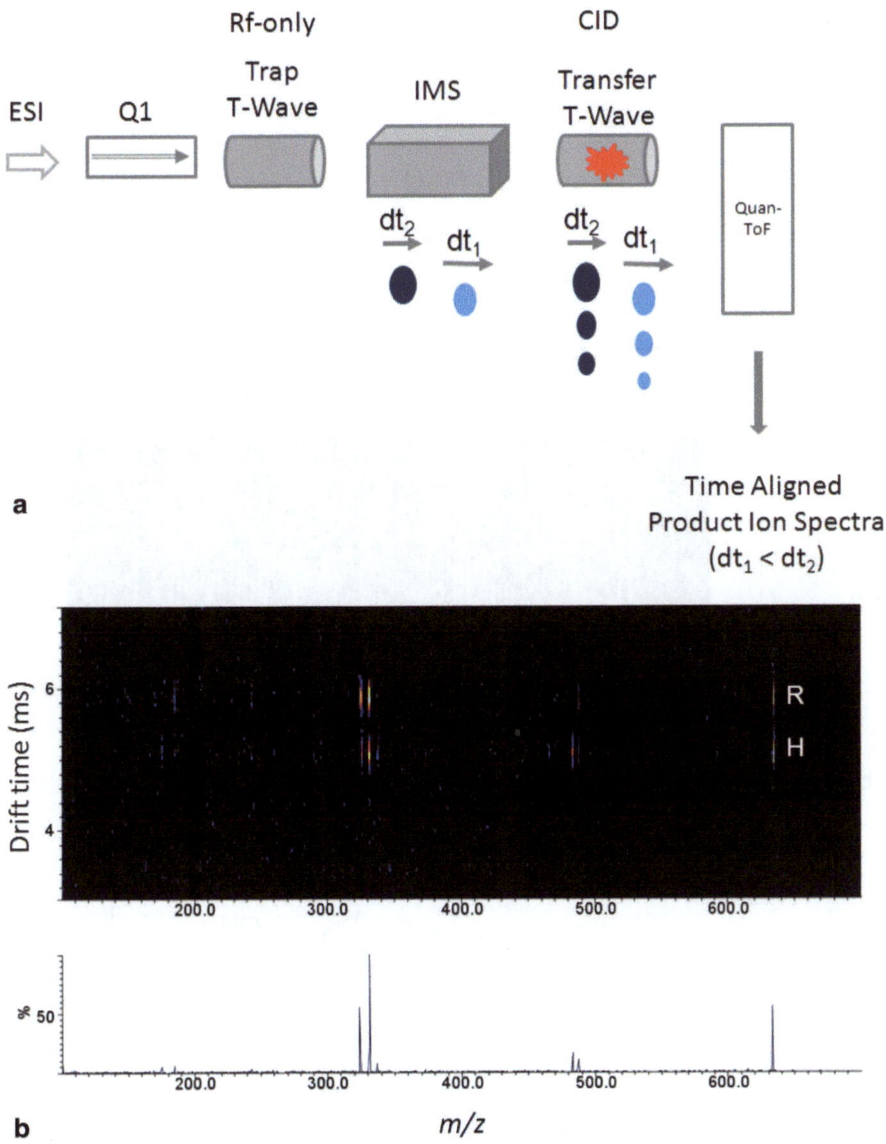

Fig. 2.4 a–b Transfer fragmentation mode for the characterization of bioactive ingredients in dietary supplements. **a** Transfer fragmentation mode—conceptual presentation. In this mode, fragment ions generated in the transfer region can be correlated to their time-aligned precursor ions. This mode of operation is particularly useful for distinguishing isobaric precursor ions as long as the precursor ions have different interaction cross sections; *v*, velocity; *dt*, drift time. **b** TWIMS–tandem mass spectrometry analysis of a supplement that contains both flavonoid glycosides, hesperidin and rutin. Transfer dissociation was conducted by selecting the ions at *m/z* 633 in the quadrupole device Q1, separation of the ions in the TWIMS region, and by collisional activation in the transfer region. *Upper panel*: drift time plot displaying the time-aligned product ions for the precursor ions with drift times 5.21 ms (hesperidin, $[M+Na]^+$) and 5.97 ms (rutin, $[M+Na]^+$). *Lower panel*: extracted tandem mass spectrum after integration of the ion signals that contribute to both time-aligned drift time distributions at 5.21 and 5.97 ms. *H*, hesperidin; *R*: rutin

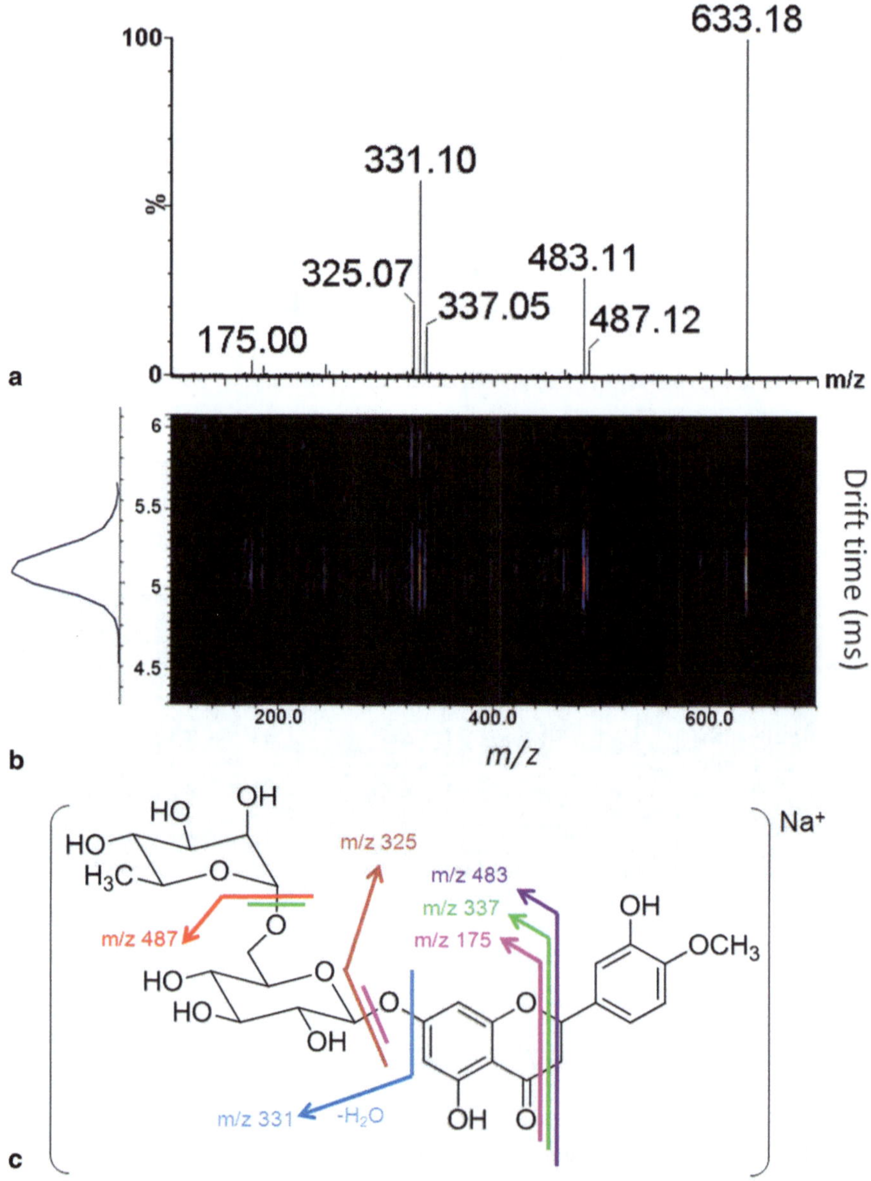

Fig. 2.5 ESI q-IMS–MS/MS transfer dissociation analysis of the sodiated hesperidin ions ($[M+Na]^+$, *m/z* 633). **a** Fragmentation spectrum, **b** drift time distribution, and **c** the proposed transfer dissociation pathways of sodiated hesperidin

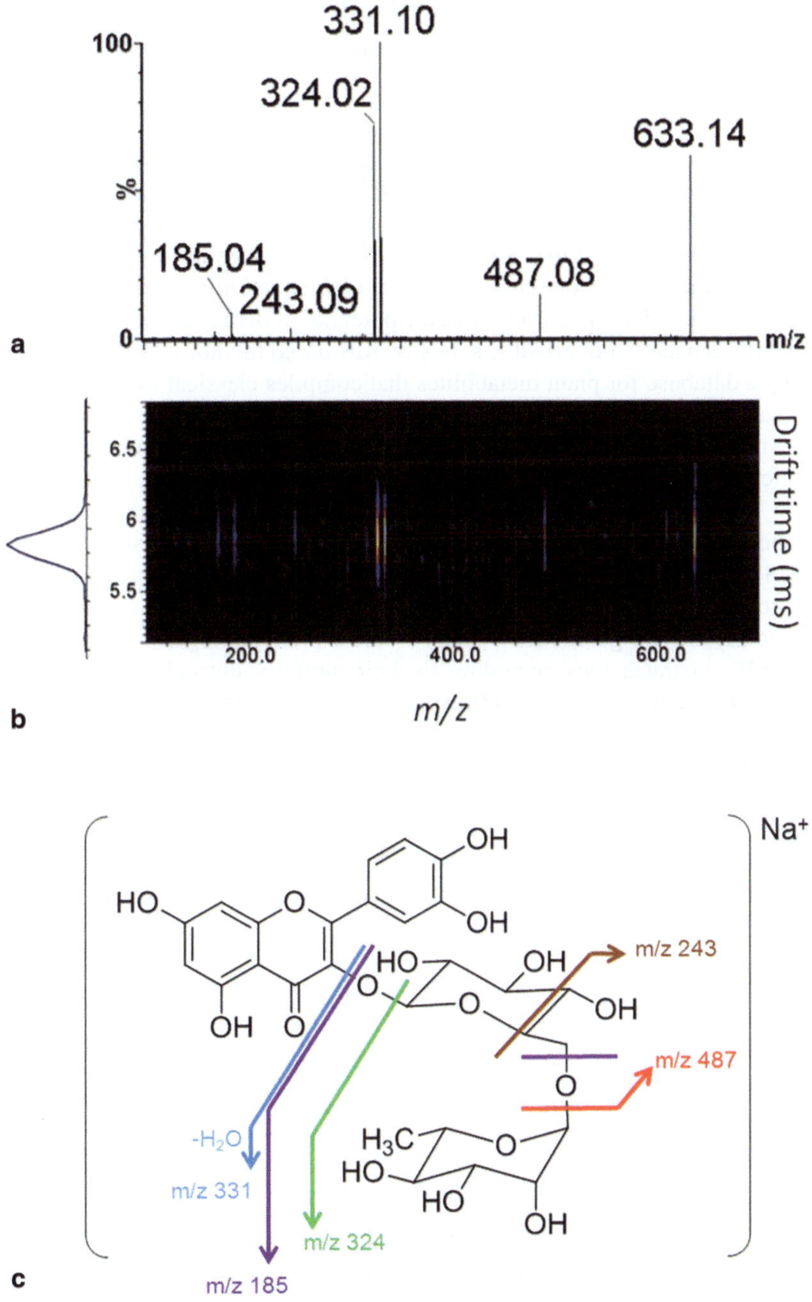

Fig. 2.6 ESI q-IMS–MS/MS transfer dissociation analysis of the sodiated rutin ions ($[M+Na]^+$, *m/z* 633). **a** Fragmentation spectrum, **b** drift time distribution, and **c** the proposed transfer dissociation pathways of sodiated rutin

2.5 TWIMS–MS for Obtaining Collision Cross Sections as an Additional Parameter for Plant Metabolite Characterization

In metabolomics, the characterization of a metabolite, e.g., a plant natural compound, is currently based on accurate mass, fragment ion spectra, and retention time comparison with a standard compound. Collision cross sections are independent of chromatographic conditions and therefore collision cross-section determinations would add an additional dimension to the characterization of a metabolite. Collision cross-section determinations have the potential of distinguishing structural isomers which is usually not possible solely by MS-based methods. We have started to build up a database for plant metabolites that compiles classical MS data (accurate mass, tandem mass spectral data) and collision cross sections. Here, we report on the determination of collision cross sections for a selection of hop phenolics. Hop phenolics have recently attracted public attention because of their health-promoting effects. These compounds show antiproliferative activity, cancer chemopreventive, and antioxidant properties [2]. Specifically, we were interested in applying TWIMS–MS for the characterization of isobaric hop phenolics, namely (i) the hop chalcone xanthohumol (XN)and the isobaric prenylated flavonoid isoxanthohumol (IX), and (ii) the two geometric isomers 6- and 8-prenylnaringenin (6-PN, 8-PN).

TWIMS separates ions according to their mobility through a continuous sequence of transient voltage pulses (traveling waves). The mobility of an ion through the T-wave cell depends on the charge on the ion, the mass of the ion and the buffer gas, the identity, temperature and pressure of the buffer gas, and its collision cross section. Hence, the collision cross section, Ω, of an ion can be expressed as given in Eq. 2.1. Since the separation of ions in the TWIMS section of the instrument is more complex than a classical drift tube, additional parameters (A and B) need to be included to account for the nonlinear effects of the TWIMS device [24, 27]:

$$\Omega = Ze\left[\frac{1}{m_1}+\frac{1}{m_N}\right]^{1/2} A t_D{}^{B} \qquad (2.1)$$

Ω = Collision cross section
Z = Number of charges on the analyte ion
e = Charge on an electron
m_1 = Mass of the analyzed ion
m_N = Mass of the buffer gas
A = Correction factor for the electric field parameters
B = Correction factor for the nonlinear effect of the TWIMS device
t_D = Drift time

Since the separation of ions in the TWIMS section of the instrument is more complex than a classical drift tube, which uses a constant electric field, the T-wave mobility separation device needs to be calibrated. For this purpose, a drift time calibration procedure was applied that uses absolute cross-section values of polyglycine (Poly-Gly) and polyalanine (Poly-Ala) peptide ions known from classical

Table 2.2 Estimated collision cross-section values of hop phenolic compounds which were obtained by TWIMS–MS

Compound	Estimated CCS[a] (Å^2)			Theoretical CCS[a], (Å^2)
	Poly-Ala[b]	Poly-Gly[b]	Poly-Ala[b] and Poly-Gly[c]	
XN[d]	128.3	128.9	125.4	127.7
IX[e]	123.5	122.8	120.4	120.3
6-PN[f]	122.9	n.d.	120.2	118.7
8-PN[g]	119.9	n.d.	117.2	122.1

Experimentally estimated cross sections were derived by comparison with poly-DL-alanine mixture (Poly-Ala) and oligo-glycine (Poly-Gly) mixture following the procedure described by Williams et al. [32]. For comparison, theoretical cross sections are listed in the right column of the table

[a] Collision cross section
[b] Poly-DL-alanine
[c] Polyglycine
[d] Xanthohumol
[e] Isoxanthohumol
[f] 6-Prenylnaringenin
[g] 8-Prenylnaringenin

drift-tube ion-mobility studies [26, 35, 36]. The experimentally estimated cross-section values were then compared to the calculated cross sections (Table 2.2). The theoretical cross sections of the compounds were obtained by using the DriftScope software (v4.1). This software uses the projection approximation (PA) approach which is also one of the models in the MOBCAL software, a program to calculate mobilities which was developed by Martin Jarrold's group [37]. Briefly, MOBCAL calculates the collision cross-sections based on three different models. These models are the exact hard sphere scattering (EHSS), projection approximation (PA), and the trajectory model™. PA calculates cross sections in regard to a collision between the buffer gas and the analyte atoms [38–40].

XN and IX (M^-, *m/z* 353.15) were detected after TWIMS–MS in negative mode with slight difference in their drift time distributions (t_{DTD} ~0.2 ms). As shown in Fig. 2.7, infusion of the IX solution resulted always in two peaks with drift times differing by 0.2 ms. The peak ratio and drift times were the same as those detected for a mixture of IX and XN. This may indicate that isomerization of IX to XN occurred in the source prior to ion-mobility separation. A similar behavior was observed for carotenoids [15]. The isomeric PNs (M^-, *m/z* 339.12) showed distinct peaks that showed only marginally different drift time distributions under the conditions used. The experimentally estimated cross sections were in good agreement with the theoretical values. This example demonstrates that TWIMS–MS is a powerful technique that is capable of providing access to an additional analytical parameter for the in-depth characterization of phytochemicals, namely experimentally estimated collision cross sections besides exact mass determination and fragment ion information. However, more research needs to be done on structurally diverse sets of small molecules to explore the full potential and limitation of TWIMS–MS.

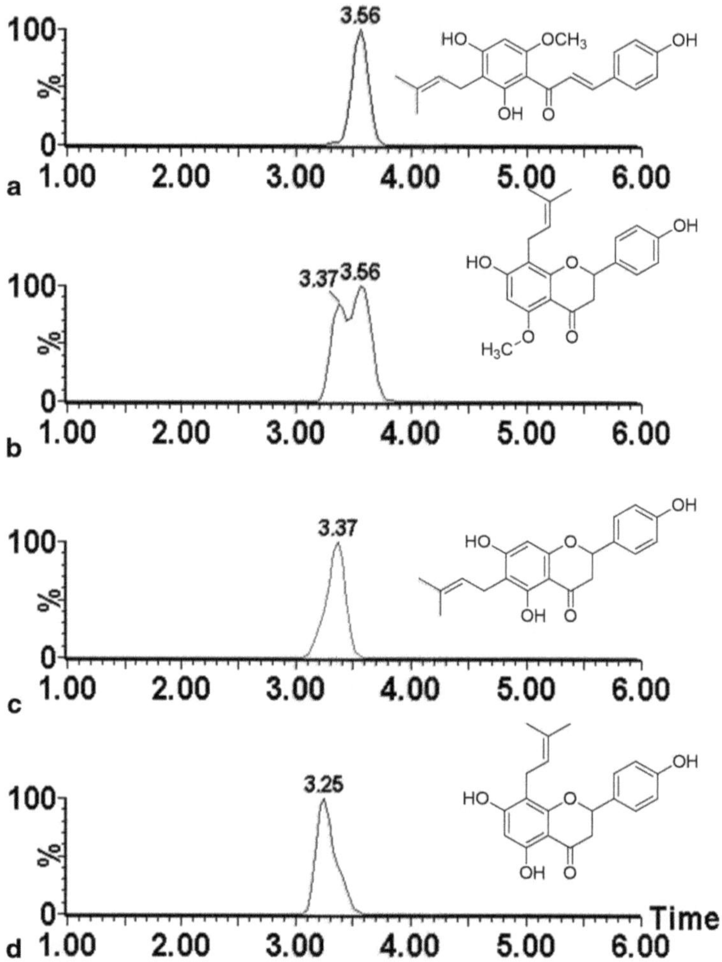

Fig. 2.7 **a–d** Drift time distributions and structures of the hop phenolics XN, IX, 6- and 8-PN. Negative ion electrospray TWIMS–MS analysis of the $[M-H]^-$ ions of the isomeric flavonoids XN and IX (both *m/z* 353.15, $C_{21}H_{21}O_5^-$) and the regioisomers 6- and 8-PN (*m/z* 339.12, $C_{20}H_{19}O_5^-$) was conducted following infusion. Drift time distributions were recorded for **a** XN at 3.56 ms, **b** IX at 3.37 and 3.56 ms, and **c** 6-PN at 3.37 ms and **d** 8-PN at 3.25 ms. Further work is needed to rationalize the possible isomerization of IX during the electrospray process. Calibration of the TWIMS device enables the estimation of the collision cross section (see Table 2.2). *XN*, xanthohumol; *IX*, isoxanthohumol; *6-PN*, 6-prenylnaringenin; *8-PN*, 8-prenylnaringenin; *time*, drift time in milliseconds (ms)

Experimental Details Stock solutions of XN, IX, 6-PN, and 8-PN were prepared separately (10 µg/mL) in methanol from solid standards. Equal amounts of XN and IX solutions (1:1 v/v) were mixed with 50 % water before use to minimize degradation and isomerization, and then immediately infused into the ESI source at a flow rate of 5 µL/min. A solution containing both prenylated naringenins, 6-PN and 8-PN, was prepared and infused. Mass calibrations in both modes (positive and negative) were done by infusing a solution of sodium formate (5 mM; 1:1, v/v, acetonitrile:water). The instrument was operated in the resolution mode. Source parameter settings used for negative ion acquisition were a capillary voltage of 2.0 kV and a sampling cone voltage of 25.0 V. The other optimized negative ion electrospray conditions were as follows: extraction cone voltage, 4.0 V; ion source temperature, 100 °C; desolvation temperature, 300 °C; desolvation gas flow rate, 500 L/h; and the cone gas flow was off. Ion-mobility separation settings were ion-mobility gas flow rate of 110 mL/min, wave velocity ramping from 500 to 570 m/s, and wave height of 40.0 V. Data acquisitions were carried out using Waters MassLynx (v4. 1) and ion-mobility spectra were processed using DriftScope software (v2.1, Waters).

The calibration of the T-wave cell was achieved by infusing a solution of polyalanine (Poly-Ala, 0.1 mg/mL, 1:1 acetonitrile:water) and a solution of six oligoglycines (Poly-Gly, 0.1 mg/mL in 1:1 acetonitrile:water). The flow rate was 5 μL/min. The electrospray source was operated in the positive mode and the settings were a capillary voltage of 2.25 V and a sampling cone voltage of 25 V. The ion-mobility separation conditions used for the calibration of the T-wave device were identical to those used for the analytes.

2.6 Conclusion and Perspectives

In summary, we discussed several applications of TWIMS–MS for the analysis of plant phenolics and demonstrated that combining ion-mobility separation and gas-phase fragmentation adds an extra level of selectivity and specificity for the structural characterization of plant metabolites.

We explored the use of ESI TWIMS–MS for the examination of highly complex biopolymers, namely grape seed proanthocyanidins. The use of ion-mobility separation prior to high-resolution accurate MS resulted in reduced spectral complexity compared to ESI–MS acquisitions alone. In particular, the presentation of TWIMS–MS data in a 2-dimensional contour plot, drift time vs. m/z, allows the deconvolution of many spectral features associated with the inherent heterogeneity of proanthocyanidin mixtures. We were able to extract proanthocyanidin oligomers that differ in their compositions. The capability of TWIMS–MS to separate proanthocyanidin ions with different charge groups enabled the analysis of proanthocyanidins with higher degree of polymerization compared to ESI–MS alone. There is an increasing interest in proanthocyanidins as dietary supplements, but the methods for characterizing proanthocyanidin preparations are sparse and mainly limited to gel permeation chromatography and acid hydrolysis in combination with liquid chromatography [30, 41]. Traditionally, MALDI MS has been the method of choice for the characterization of proanthocyanidin oligomers and higher-molecular-weight tannins [28, 30]. ESI–MS, in particular in combination with liquid chromatography, shows promise for the characterization of proanthocyanidins. Its broad use has been hampered by the limitation that the extraction of oligomer distribution information is complicated due to overlapping charge state distributions of the different oligomers [30, 33]. The possible inclusion of TWIMS may provide the means to advance the analysis of proanthocyanidins. It would be interesting to evaluate MALDI TWIMS–MS for the analysis of proanthocyanidins, as it is likely that this will further improve the analysis of this class of compounds. The additional insight into the structural complexity of proanthocyanidin preparations may ultimately lead to standardized proanthocyanidin preparations for research purposes and the dietary supplements market. Beyond the current application to proanthocyanidin oligomers, IMS–MS has been used for the characterization of synthetic polymers [11, 12]. We foresee that TWIMS–MS and alternative IMS–MS techniques will significantly impact the way we characterize other biopolymers and bio-inspired polymeric materials.

TWIMS–MS technology provides access to drift time information and experimentally estimated collision cross sections, analytical parameters that are not available on MS/MS-only instruments. The potential separation and assignment of positional isomers of natural products and metabolites is an emerging application of TWIMS–MS. The potential of TWIMS of distinguishing structural isomers that differ in the site of substitutions has been proven to be a particularly powerful technique for drug metabolism studies [13, 14]. The combination of molecular modeling studies for deriving theoretically derived cross sections with TWIMS measurements of cross sections provides an exciting new strategy for the assignment of metabolite isomers. Here, an ensemble of energy-minimized metabolite structures is generated "in silico." For the ensemble of energy-minimized structures, the theoretical cross sections are calculated. The experimentally derived cross sections are then compared with the theoretical cross sections to support the assignment of discrete metabolite structures. For instance, this approach has been successfully demonstrated for the elucidation of hydroxylation sites of ondansetron metabolites [13]. In the ondansetron metabolite study, TWIMS–MS was capable of reproducibly measuring drift time distribution differences between metabolites of 20 μs corresponding to a difference of only 0.3 Å in experimentally derived cross sections. Similarly, the capability of measuring small differences in drift times (50 μs corresponding to 0.6 Å) with TWIMS–MS was also shown for organoruthenium anticancer complexes where the arene is ortho- or meta-terphenyl [35]. Due to the increasing interest in applying TWIMS–MS characterization to small molecule studies, a small database of TWIMS-derived and validated collision cross sections for pharmaceutically relevant compounds has become available recently, which will certainly further broaden the use of TWIMS–MS for small molecule analyses [42]. Since the advent of the first commercially available TWIMS–MS instrument in 2006, technology has advanced significantly resulting in increased resolution of the traveling-wave ion-mobility separator [25], new applications [19, 43–47], and much enthusiasm in the scientific community. We foresee that TWIMS–MS and alternative IMS–MS techniques will significantly impact and advance our measuring capabilities in the phytochemical and pharmaceutical sciences, and in the emerging field of metabolomics [17, 18, 48, 49].

Advancements in system biology research will depend on tools that are capable of deconvoluting highly complex systems. The analysis of complex systems requires the availability of a new generation of separation and measurement tools. Recent applications of IMS–MS to a broad range of biomolecule measurements have demonstrated that IMS–MS has emerged as a powerful analytical technique capable of providing the separation space necessary to analyze highly complex samples [17, 23]. The combination of devices that allow real-time monitoring of living systems using IMS–MS is another exciting avenue of facilitating system-biology experiments [50]. The future for IMS–MS is bright and full of opportunities.

References

1. Tsao R (2010) Chemistry and biochemistry of dietary polyphenols. Nutrients 2:1231–1246
2. Stevens JF, Page JE (2004) Xanthohumol and related prenylflavonoids from hops and beer: to your good health! Phytochemistry 65:1317–1330
3. Miranda CL et al (2012) Flavonoids. eLS, Wiley Online Library
4. Counterman AE, Clemmer DE (2002) Cis-trans signatures of proline-containing tryptic peptides in the gas phase. Anal Chem 74:1946–1951
5. Kaleta DT, Jarrold MF (2003) Helix-turn-helix motifs in unsolvated peptides. J Am Chem Soc 125:7186–7187
6. Ruotolo BT et al (2004) Ion mobility-mass spectrometry applied to cyclic peptide analysis: conformational preferences of gramicidin S and linear analogs in the gas phase. J Am Soc Mass Spectrom 15:870–878
7. Ruotolo BT et al (2002) Observation of conserved solution-phase secondary structure in gas-phase tryptic peptides. J Am Chem Soc 124:4214–4215
8. Ruotolo BT et al (2005) Evidence for macromolecular protein rings in the absence of bulk water. Science 310:1658–1661
9. Uetrecht C et al (2010) Ion mobility mass spectrometry of proteins and protein assemblies. Chem Soc Rev 39:1633–1655
10. Zilch LW et al (2007) Folding and unfolding of helix-turn-helix motifs in the gas phase. J Am Soc Mass Spectrom 18:1239–1248
11. Trimpin S, Clemmer DE (2008) Ion mobility spectrometry/mass spectrometry snapshots for assessing the molecular compositions of complex polymeric systems. Anal Chem 80:9073–9083
12. Trimpin S et al (2007) Resolving oligomers from fully grown polymers with IMS-MS. Anal Chem 79:7965–7974
13. Dear GJ et al (2010) Sites of metabolic substitution: investigating metabolite structures utilising ion mobility and molecular modelling. Rapid Commun Mass Spectrom 24:3157–3162
14. Cuyckens F et al (2011) Product ion mobility as a promising tool for assignment of positional isomers of drug metabolites. Rapid Commun Mass Spectrom 25:3497–3503
15. Dong L et al (2010) Collision cross-section determination and tandem mass spectrometric analysis of isomeric carotenoids using electrospray ion mobility time-of-flight mass spectrometry. Anal Chem [Epub ahead of print]
16. Bohrer BC, Clemmer DE (2011) Biologically-inspired peptide reagents for enhancing IMS-MS analysis of carbohydrates. J Am Soc Mass Spectrom 22:1602–1609
17. Dwivedi P et al (2010) Metabolic profiling of human blood by high resolution ion mobility mass spectrometry (IM-MS). Int J Mass Spectrom 298:78–90

18. Castro-Perez J et al (2011) Localization of fatty acyl and double bond positions in phosphatidylcholines using a dual stage CID fragmentation coupled with ion mobility mass spectrometry. J Am Soc Mass Spectrom 22:1552–1567
19. Kliman M et al (2011) Lipid analysis and lipidomics by structurally selective ion mobility-mass spectrometry. Biochim Biophys Acta 1811:935–945
20. Kanu AB et al (2008) Ion mobility-mass spectrometry. J Mass Spectrom 43:1–22
21. Verbeck GF et al (2002) A fundamental introduction to ion mobility mass spectrometry applied to the analysis of biomolecules. J Biomol Tech 13:56–61
22. Bohrer BC et al (2008) Biomolecule analysis by ion mobility spectrometry. Annu Rev Anal Chem (Palo Alto Calif) 1:293–327
23. Liu X et al (2007) Mapping the human plasma proteome by SCX-LC-IMS-MS. J Am Soc Mass Spectrom 18:1249–1264
24. Giles K et al (2004) Applications of a travelling wave-based radio-frequency-only stacked ring ion guide. Rapid Commun Mass Spectrom 18:2401–2414
25. Giles K et al (2011) Enhancements in travelling wave ion mobility resolution. Rapid Commun Mass Spectrom 25:1559–1566
26. Michaelevski I et al (2010) T-wave ion mobility-mass spectrometry: basic experimental procedures for protein complex analysis. J Vis Exp 41: e1985
27. Smith DP KT, Campuzano I, Malham RW, Berryman JT, Radford SE, Ashcroft AE (2009) Deciphering drift time measurements from travelling wave ion mobility spectrometry-mass spectrometry studies. Eur J Mass Spectrom (Chichester Eng) 15:113–130
28. Monagas M et al (2010) MALDI-TOF MS analysis of plant proanthocyanidins. J Pharm Biomed Anal 51:358–372
29. Mouls L et al (2011) Comprehensive study of condensed tannins by ESI mass spectrometry: average degree of polymerisation and polymer distribution determination from mass spectra. Anal Bioanal Chem 400:613–623
30. Taylor AW et al (2003) Hop (*Humulus lupulus* L.) proanthocyanidins characterized by mass spectrometry, acid catalysis, and gel permeation chromatography. J Agric Food Chem 51:4101–4110
31. Porter PJ (1988) Flavans and proanthocyanidins. The Flavonoids 21–62
32. Aron PM, Kennedy JA (2008) Flavan-3-ols: nature, occurrence and biological activity. Mol Nutr Food Res 52:79–104
33. Hayasaka Y et al (2003) Characterization of proanthocyanidins in grape seeds using electrospray mass spectrometry. Rapid Commun Mass Spectrom 17:9–16
34. Harry EL et al (2009) Direct analysis of pharmaceutical formulations from non-bonded reversed-phase thin-layer chromatography plates by desorption electrospray ionisation ion mobility mass spectrometry. Rapid Commun Mass Spectrom 23:2597–2604
35. Williams JP et al (2009) Isomer separation and gas-phase configurations of organoruthenium anticancer complexes: ion mobility mass spectrometry and modeling. J Am Soc Mass Spectrom 20:1119–1122
36. http://www.indiana.edu/~clemmer/Research/Cross%20Section%20Database/Peptides/polyaminoacid_cs.htm. Accessed 14 Oct 2013
37. http://www.indiana.edu/~nano/index.html. Accessed 14 Oct 2013
38. Wyttenbach T et al (1997) Effect of the long-range potential on ion mobility measurements. J Am Soc Mass Spectrom 8:275–282
39. Ruotolo BT et al (2008) Ion mobility-mass spectrometry analysis of large protein complexes. Nat Protoc 3:1139–1152
40. Knapman TW et al. (2010) Determining the topology of virus assembly intermediates using ion mobility spectrometry-mass spectrometry. Rapid Commun Mass Spectrom 24:3033–3042
41. Kennedy JA, Jones GP (2001) Analysis of proanthocyanidin cleavage products following acid-catalysis in the presence of excess phloroglucinol. J Agric Food Chem 49:1740–1746
42. Salbo R et al (2012) Traveling-wave ion mobility mass spectrometry of protein complexes: accurate calibrated collision cross-sections of human insulin oligomers. Rapid Commun Mass Spectrom 26:1181–1193

43. Li H et al (2012) Resolving structural isomers of monosaccharide methyl glycosides using drift tube and traveling wave ion mobility mass spectrometry. Anal Chem 84:3231–3239
44. Goodwin CR et al (2012) Structural mass spectrometry: rapid methods for separation and analysis of peptide natural products. J Nat Prod 75:48–53
45. Rand K et al (2011) ETD in a traveling wave ion guide at tuned Z-spray ion source conditions allows for site-specific hydrogen/deuterium exchange measurements. J Am Soc Mass Spectrom 22:1784–1793
46. Halgand F et al (2011) Dividing to unveil protein microheterogeneities: traveling wave ion mobility study. Anal Chem 83:7306–7315
47. Ridenour WB et al (2010) Structural characterization of phospholipids and peptides directly from tissue sections by MALDI traveling-wave ion mobility-mass spectrometry. Anal Chem 82:1881–1889
48. Dwivedi P et al (2010) Metabolic profiling of *Escherichia coli* by ion mobility-mass spectrometry with MALDI ion source. J Mass Spectrom 45:1383–1393
49. Kaplan K et al (2009) Monitoring dynamic changes in lymph metabolome of fasting and fed rats by electrospray ionization-ion mobility mass spectrometry (ESI-IMMS). Anal Chem 81:7944–7953
50. Enders JR et al (2010) Towards monitoring real-time cellular response using an integrated microfluidics-matrix assisted laser desorption ionisation/nanoelectrospray ionisation-ion mobility-mass spectrometry platform. IET Systems Biology 4:416–427

Chapter 3
Camptothecin Production and Biosynthesis in Plant Cell Cultures

Takashi Asano, Kazuki Saito and Mami Yamazaki

Abstract Camptothecin, a well-known monoterpenoid indole alkaloid originally identified in the extracts of the Chinese tree *Camptotheca acuminata* (Nyssaceae), exhibits antitumor activity due to its ability to kill cancer cells via topoisomerase I poisoning. Other plant species have since been shown to produce camptothecin and related compounds. In particular, *Ophiorrhiza* species (Rubiaceae) are important resources for the production of various alkaloids, including camptothecin. This chapter describes the production of camptothecin-related alkaloids and the elucidation of the mechanisms of camptothecin biosynthesis using plant cell and tissue cultures. In particular, aseptically grown plants, callus cultures, and hairy root cultures were established for several species, *O. liukiuensis, O. kuroiwai*, and *O. pumila*, which were then evaluated for production of camptothecin and related alkaloids. The metabolite profiles differed between the species, and between tissues of the same species; for example, profiles from hairy roots were not identical to those of aseptic plants. The complementary DNAs (cDNAs) for strictosidine synthase, tryptophan decarboxylase, and cytochrome P450 reductase were cloned from *O. pumila* and evaluated for involvement in production of camptothecin in this species. RNA interference (RNAi)-mediated knockdown of gene expression indicated that

M. Yamazaki (✉) · T. Asano · K. Saito
Graduate School of Pharmaceutical Sciences, Chiba University, Inohana 1-8-1,
Chuo-ku, Chiba 260-8675, Japan
e-mail: mamiy@faculty.chiba-u.jp

T. Asano · M. Yamazaki
CREST, Japan Science and Technology Agency, 4-1-8 Honcho, Kawaguchi,
Saitama 332-0012, Japan

K. Saito
RIKEN Center for Sustainable Resource Science, 1-7-22 Suehiro-cho, Tsurumi-ku,
Yokohama 230-0045, Japan

T. Asano
School of Pharmacy, Iwate Medical University, 2-1-1 Nishitokuta, Yahaba,
Iwate 028-3694, Japan

D. R. Gang (ed.), *50 Years of Phytochemistry Research*,
Recent Advances in Phytochemistry 43, DOI 10.1007/978-3-319-00581-2_3,

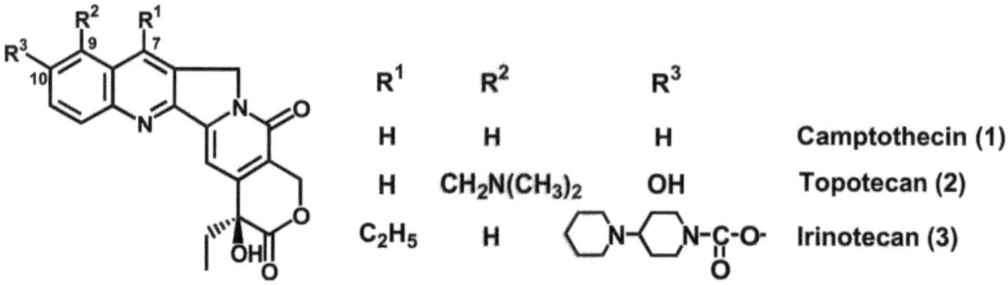

Fig. 3.1 Camptothecin (*1*) and its clinically used derivatives, topotecan (*2*) and irinotecan (*3*). (With permission from Ref. [38])

the production of camptothecin, strictosidine, and camptothecin-related alkaloids was suppressed in a *TDC* expression-dependent manner in RNAi hairy roots.

3.1 Introduction

Alkaloids are nitrogen-containing basic compounds known from about 20% of all plant species. Many alkaloids are pharmacologically active and have been used traditionally in the form of medicinal plant extracts as treatments for various diseases [1]. A few dozen pharmacologically active alkaloids, including camptothecin, are widely used in modern medicine, and worldwide sales of alkaloid-containing drugs were projected to exceed US$ 4 billion in 2002 [2].

Camptothecin (1) is a well-known monoterpenoid indole alkaloidand was originally identified in the extracts of the Chinese tree *Camptotheca acuminata* (Nyssaceae) [3]. Camptothecin exhibits antitumor activity, which is due to its ability to kill cancer cells via topoisomerase I poisoning [4]. At present, the semi-synthetic water-soluble camptothecin derivatives, topotecan (2) and irinotecan (3), are used worldwide as clinical antitumor agents against cancers of the lung, cervix, ovaries, colon [5], and other organs [6] (Fig. 3.1). In addition, a number of reports are available announcing the therapeutic values of camptothecin derivatives against acquired immunodeficiency syndrome (AIDS) [7] and falciparum malaria [8]. Consequently, the demand for camptothecin will continue to increase in the future.

Despite the rapid growth of the pharmaceutical market for this compound, camptothecin is still supplied exclusively from intact plants, mainly *C. acuminata* and *Nothapodytes foetida* [9]. However, the extraction of this compound from intact plants is problematic because of the shortage of natural resources and the resultant environmental concerns. Thus, the production of secondary metabolites by genetically engineered plant cell cultures, particularly for compounds such as camptothecin, has become a keen issue [10].

Camptothecin-related alkaloids have been reported to be produced in a relatively wide array of plant species, besides *C. acuminata* and *N. foetida* [11]. For instance,

Fig. 3.2 The genus *Ophiorrhiza* species distributed in Japan

Merrilliodendron megacarpum [12], *Pyrenacantha klaineana* (Icacinaceae) [13], *Ervatamia heyneana* (Apocynaceae) [14], *Mostuea brunonis* (Loganiaceae) [15], *Ophiorrhiza mungos* [16], and *O. filistipula* (Rubiaceae) [17] have been reported to produce camptothecin-related compounds. Moreover, the results of phytochemical studies of the genus *Ophiorrhiza* have shown that camptothecin also accumulates in some *Ophiorrhiza* species (e.g., *O. pumila*) distributed in Japan [18, 19].

The genus *Ophiorrhiza* is widely distributed around tropical and subtropical Asia and comprises about 150 species [20]. Moreover, some of these species produce indole alkaloids [21]. With regard to the chemical constituents of *Ophiorrhiza* species distributed in Japan, *O. pumila* accumulated camptothecin and related alkaloids [18, 22] and *O. japonica* accumulated β-carboline-type alkaloids, such as lyalosidic acid and harman [23, 24]. Meanwhile, *O. liukiuensis* [25] and *O. kuroiwai* [26], which was shown to be an interspecies hybrid of *O. pumila* and *O. liukiuensis*, accumulated both camptothecin-related alkaloids and β-carboline-type alkaloids (Fig. 3.2). Therefore, these *Ophiorrhiza* species are important as resources for the production of various alkaloids, including camptothecin.

In this chapter, we describe the production of camptothecin-related alkaloids and the elucidation of the mechanisms of camptothecin biosynthesis by use of plant cell and tissue cultures.

3.2 In Vitro Cultures of Camptothecin-Producing Plants

3.2.1 Establishment of In Vitro Cultures

Cell and tissue cultures of several camptothecin-producing plants have been investigated as alternative sources for camptothecin production [27]. Sakato et al. [28] reported the first establishment of a rapidly growing cell suspension culture of *C. acuminata*, although the camptothecin productivity was insufficient (0.002 mg g^{-1} dry weight) for practical use. Callus cultures of *C. acuminata* established by Wiedenfeld et al. [29] produced comparatively adequate amounts of camptothecin (2 mg g^{-1} dry weight). These callus cultures were also reported to contain 10-hydroxycamptothecin, from trace amounts up to 0.08–0.1 mg g^{-1} dry weight [29]. Callus cultures of *N. foetida* were found to accumulate small amounts of camptothecin and 9-methoxycamptothecin [30–32], but the level of alkaloid production was 100- to 1000-fold lower than that from soil-grown plants. Callus cultures of *O. pumila* produced no camptothecin-related alkaloids but accumulated only anthraquinones [33, 34].

Since alkaloid biosynthesis and accumulation are under the strict control of cell developmental and environmental factors [35], establishing cultures of cell types suitable for the production of the camptothecin is important. Accordingly, aseptic plants and hairy roots of *Ophiorrhiza* species have been established as an effective means of producing camptothecin (Fig. 3.3) [36–38].

3.2.2 Camptothecin Production and Metabolite Profiles in Tissue Cultures of Ophiorrhiza Species

In shoots and roots of established aseptic plants of *Ophiorrhiza* species, camptothecin production per tissue weight was the highest in the roots of *O. pumila*. On the other hand, the production per tube was the highest in *O. kuroiwai* because it showed the higher growth rate of the two species. The concentration and total amount of camptothecin in *O. liukiuensis* were lower than those of *O. kuroiwai* and *O. pumila*.

Camptothecin accumulated to higher levels in hairy root lines of *O. pumila* than in those of *O. liukiuensis* and *O. kuroiwai* [38]. Camptothecin accumulation and increased growth rate of *O. pumila* hairy roots have the best results in the reports of camptothecin production by *in vitro* tissue cultures [37, 39].

The patterns of secondary metabolite production in the aseptic plants and hairy roots of *Ophiorrhiza* species were profiled by high-performance liquid chromatography–diode array detection–electrospray ion trap tandem mass spectrometry (Fig. 3.4 and Table 3.1) [38]. The metabolite profiles of *O. liukiuensis* and *O. kuroiwai* were highly similar in the shoot and root. 10-Methoxycamptothecin (5) and lyalosidic acid (6) were detected in the roots and shoots, respectively, of both *O. liukiuensis* and *O. kuroiwai* but not in those of *O. pumila*. Moreover, 3(*S*)- and

Fig. 3.3 Established tissue cultures of *Ophiorrhiza liukiuensis, O. kuroiwai*, and *O. pumila*. **a** Aseptic plants cultured for 5 weeks on 1/2 MS medium containing 1 % sucrose and 0.2 % gellan gum in test tubes. **b** Hairy roots cultured for 4 weeks in B5 liquid medium containing 2 % sucrose. (With permission from Ref. [38])

3(*R*)-deoxypumilosides (9, 10) were detected only in *O. pumila*. Camptothecin (1), 9-methoxycamptothecin (4), strictosamide (7), pumiloside (8), strictosidinic acid (11), and 3-*O*-caffeoylquinic acid (13) were detected in all three species. The metabolite profiles of the hairy roots were not identical to those of aseptic plants.

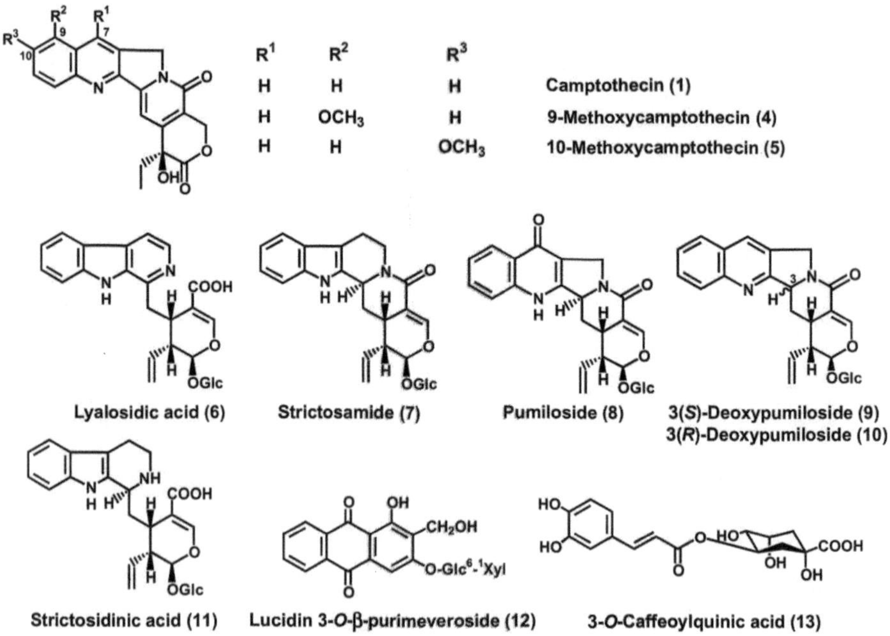

Fig. 3.4 Chemical structures of secondary metabolites detected in tissue cultures of *Ophiorrhiza* species. (With permission from Ref. [38])

Table 3.1 Alkaloids and anthraquinones detected in tissue cultures of *Ophiorrhiza* species

Compound	*O. liukiuensis*			*O. kuroiwai*			*O. pumila*		
	Shoot	Root	Hairy root	Shoot	Root	Hairy root	Shoot	Root	Hairy root
1 Camptothecin	+	+	+	+	+	+	+	+	+
4 9-Methoxycamptothecin	+	+		+	+		+	+	
5 10-Methoxycamptothecin		+			+				
6 Lyalosidic acid	+			+					
7 Strictosamide		+	+		+	+	+	+	
8 Pumiloside	+	+	+	+	+	+	+	+	+
9 3(S)-Deoxypumiloside[a]							+	+	
10 3(R)-Deoxypumiloside[a]									
11 Strictosidinic acid	+			+			+		
12 Lucidin 3-O-β-purimeveroside									+
13 3-O-Caffeoylquinic acid	+			+	+		+	+	+

[a] 3(*S*)-Deoxypumiloside (9) and 3(*R*)-deoxypumiloside (10) cannot be separated in this condition

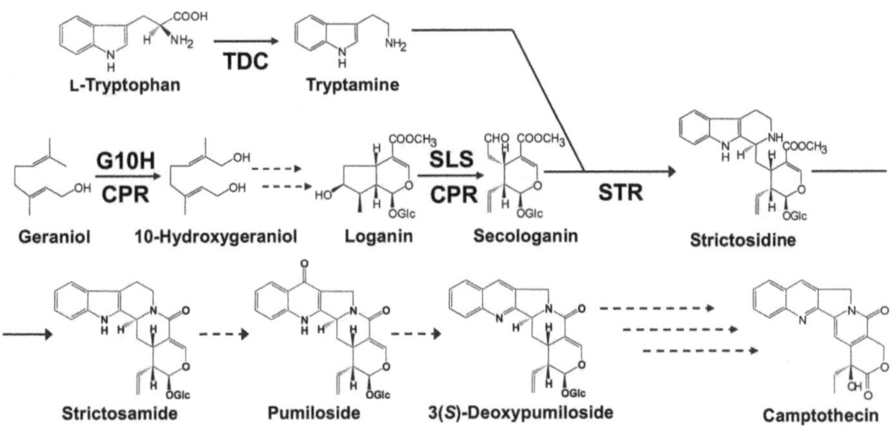

Fig. 3.5 Predicted camptothecin biosynthetic pathway in *O. pumila*. The enzymes are as follows: *TDC*, tryptophan decarboxylase; *G10H*, geraniol 10-hydroxylase; *CPR*, NADPH:cytochrome *P450* reductase; *SLS*, secologanin synthase; *STR*, strictosidine synthase. Plausible intermediates of camptothecin biosynthesis are provided in *parentheses*

3.3 Biosynthesis of Camptothecin

3.3.1 Camptothecin Biosynthetic Genes

Monoterpenoid indole alkaloids, including camptothecin, are derived from strictosidine, which is a common intermediate formed by condensation of the indole tryptamine with the iridoid glucoside secologanin by the enzyme strictosidine synthase (STR) [40–42] (Fig. 3.5). The intramolecular cyclization of strictosidine results in strictosamide, which is an intermediate peculiar to camptothecin biosynthesis, as proven by the incorporation of radiolabeled precursor [43]. The remaining details between strictosamide and camptothecin are not completely defined. However, camptothecin has been postulated to be formed potentially from strictosamide by three transformations: (1) oxidation–recyclization of the B- and C-rings, (2) oxidation of the D-ring and removal of the C-21 glucose moiety, and (3) oxidation of ring E [43]. Plausible camptothecin precursors, such as pumiloside and 3(*S*)-deoxypumiloside, were isolated from *Ophiorrhiza* species [18, 19]. Pumiloside has been found also in *C. acuminata* [44].

The cloning of complementary DNAs (cDNAs) from *O. pumila* hairy roots has been successfully performed to isolate the genes encoding the biosynthetic enzymes involved in the upper part of camptothecin biosynthesis, including STR, tryptophan decarboxylase (TDC) [45], and nicotinamide adenine dinucleotide phosphate, reduced form (NADPH):cytochrome P450 reductase (CPR), in this species [46] (Fig. 3.5). The full-length STR cDNA sequence isolated from *O. pumila* (*OpSTR*) contained a 1,056-bp open reading frame (ORF) encoding a protein of 351 amino acids with a molecular mass of 38.9 kDa. The deduced amino acid sequence of

OpSTR exhibited 55 % and 51 %identities with STRs from *Rauwolfia serpentina* [41] and *Catharanthus roseus* [47], respectively. OpSTR most likely localizes to the vacuole, as predicted by the PSORT program. Southern blot analysis suggested that a single STR-encoding gene is present in the genome of *O. pumila*. The highest *OpSTR* expression occurred in hairy roots, followed by the root, and the stem, whereas *OpSTR* was apparently not expressed in leaves. STR enzymatic activity was detected in the protein extracts of stems, roots, and hairy roots; however, no activity was detected in leaf and callus extracts. The distribution of STR activity correlated with the messenger RNA (mRNA) accumulation pattern and the camptothecin concentrations in *O. pumila* tissues, with the exception of the young leaves, suggesting that roots and stems are the main tissues for camptothecin biosynthesis [34].

Tryptamine, a precursor of strictosidine, is formed by the decarboxylation of tryptophan by the enzyme TDC. The cDNA clone encoding TDC was first isolated from *C. roseus* [48]. The full-length TDC cDNA sequence isolated from *O. pumila* (*OpTDC*) contained a 1,521-bp ORF encoding a protein of 506 amino acids with a molecular mass of 56.6 kDa. The deduced amino acid sequence of OpTDC showed high identity to TDCs from *C. acuminata* [49] and *C. roseus* [48] (71 and 67 %, respectively). Southern blot analysis suggested that at least TDC-encoding genes are present in the genome. The expression patterns of *OpSTR* and *OpTDC* were nearly the same.

The enzyme CPR is essential for the activity of cytochrome P450 monooxygenases, such as geraniol 10-hydroxylase (G10H) and secologanin synthase (SLS), which are involved in camptothecin biosynthesis [50] (Fig. 3.5). The full-length CPR cDNA sequence isolated from *O. pumila* (*OpCPR*) contained a 2,073-bp ORF encoding a protein of 690 amino acids with a molecular mass of 76.6 kDa. The deduced amino acid sequence of OpCPR showed high identity with *Arabidopsis thaliana, Petroselinum crispum*, *Pisum sativum*, and *Triticum aestivum* CPRs (72, 66, 65, and 67 %, respectively). Southern blot analysis suggested that only a single CPR-encoding gene was present in the genome of *O. pumila*. Mirroring the general importance of the enzyme, *OpCPR* was expressed in all tissues.

Studies have been performed to investigate the effects of stress compounds, such as methyl jasmonate (MeJA), salicylic acid (SA), and yeast extract (YE), on the expression of *OpSTR*, *OpTDC*, and *OpCPR* in *O. pumila* hairy roots [46]. The changes in the expression patterns of *OpSTR* and *OpTDC* in response to these various compounds were highly similar. In particular, *OpSTR* and *OpTDC* expression was repressed by SA and YE treatments but unaffected by MeJA. Meanwhile, no treatment resulted in the induction or repression of *OpCPR* transcripts. In addition, no change in STR activity was observed after treatment with either stress compounds or phytohormones.

3.3.2 In Silico and In Vitro Tracer Studies with [1-13C] glucose

Both the mevalonate (MVA) pathway [51] and the 2*C*-methyl-D-erythritol 4-phosphate (MEP) pathway [52–54] have been recognized for their role in the formation of isopentenyl diphosphate, the precursor of terpenoid biosynthesis. Yamazaki et al. [55] investigated the incorporation of [1-^{13}C]glucose into camptothecin in the hairy roots of *O. pumila* by *in silico* computation using the Atomic Reconstruction of Metabolism (ARM) [56] program and by *in vivo* tracer experiments. The ^{13}C-nuclear magnetic resonance (^{13}C-NMR) analysis clearly showed that the secologanin moiety of camptothecin was synthesized via the MEP pathway. Furthermore, in *O. pumila* hairy root cultures, treatment with fosmidomycin, a specific inhibitor of the MEP pathway, resulted in a significant decrease in camptothecin production. These results support the conclusion that the secologanin moiety of camptothecin is derived from the MEP pathway.

3.4 Metabolic Modification in Hairy Roots of *O. pumila* by RNA Interference

A detailed understanding of camptothecin production, including the enzymatic pathway for its biosynthesis, will be essential to the ultimate goal of the metabolic engineering of this compound. In *Papaver somniferum* (opium poppy), genetic approaches using antisense RNA [57, 58] or RNA interference (RNAi)-mediated silencing [59] of biosynthetic enzymes have been performed, leading to rapid progress in the metabolic engineering of benzylisoquinoline alkaloids. Therefore, it is considered that RNAi technology is an effective strategy for investigating camptothecin biosynthesis. In our study, the production of camptothecin, strictosidine, and camptothecin-related alkaloids was suppressed in a *TDC* expression-dependent manner in RNAi hairy roots. Among the hairy root-specific peaks correlated with *TDC* expression in the liquid chromatography/Fourier transform ion cyclotron resonance mass spectrometry (LC-FTICR-MS) analysis, two unknown peaks with a positive correlation were annotated as alkaloids and six unknown peaks with a negative correlation, as flavonoids. The exact mass of several non-annotated peaks was similar to those of predicted intermediates in camptothecin biosynthesis, suggesting that most peaks that positively correlated with *TDC* expression could be intermediates in camptothecin biosynthesis [60].

Acknowledgments This work was supported in part by the Grant-in-Aid for Scientific Research on Innovative Areas from The Ministry of Education, Culture, Sports, Science and Technology (MEXT), and by CREST of the Japan Science and Technology Agency (JST).

References

1. Kutchan TM (1995) Alkaloid biosynthesis: the basis for metabolic engineering of medicinal plants. Plant Cell 7:1059–1070
2. Raskin I, Ribnicky DM, Momarnytsky S et al (2002) Plants and human health in the twenty-first century. Trends Biotechnol 20:522–531

3. Wall ME, Wani MC, Cook CE et al (1966) Plant antitumor agents. I. The isolation and structure of camptothecin, a novel alkaloidal leukemia and tumor inhibitor from *Camptotheca acuminata*. J Am Chem Soc 88:3888–3890
4. Hsiang YH, Hertzberg R, Hecht S et al (1985) Camptothecin induces proteinlinked DNA breaks via mammalian DNA topoisomerase I. J Biol Chem 260:14873–14878
5. Giovanella BC, Stehlin JS, Wall ME et al (1989) DNA topoisomerase I-targeted chemotherapy of human colon cancer in xenografts. Science 246:1046–1048
6. Pizzolato JF, Saltz LB (2003) The camptothecins. Lancet 361:2235–2242
7. Priel E, Showalter SD, Blair DG (1991) Inhibition of human immunodeficiency virus (HIV-1) replication in vitro by noncytotoxic doses of camptothecin, a topoisomerase I inhibitor. AIDS Res Hum Retroviruses 7:65–72
8. Bodley AL, Cumming JN, Shapiro TA (1998) Effects of camptothecin, a topoisomerase I inhibitor, on *Plasmodium falciparum*. Biochem Pharmacol 55:709–711
9. Govindachari TR, Viswanathan N (1972) Alkaloids of *Mappia foetida*. Phytochemistry 11:3529–3531
10. Springob K, Saito K (2002) Metabolic engineering of plant secondary metabolism: a promising approach to the production of pharmaceuticals. Sci Cult 68:76–85
11. Lorence A, Nessler CL (2004) Camptothecin, over four decades of surprising findings. Phytochemistry 65:2735–2749
12. Arisawa M, Gunasekera SP, Cordell GA et al (1981) Plant anticancer agents XXI. Constituents of *Merrilliodendron megacarpum*. Planta Med 43:404–407
13. Zhou B-N, Hoch JM, Johnson RK et al (2000) Use of COMPARE analysis to discover new natural product drugs: isolation of camptothecin and 9-methoxycamptothecin from a new source. J Nat Prod 63:1273–1276
14. Gunasekera SP, Badawi MM, Cordell GA et al (1979) Plant anticancer agents X. Isolation of camptothecin and 9-methoxycamptothecin from *Ervatamia heyneana*. J Nat Prod 42:475–477
15. Dai J-R, Halloch YF, Cardellina II JH et al (1999) 20-*O*-*β*-Glucopyranosyl camptothecin from *Mostuea brunonis*: a potential camptothecin pro-drug with improved solubility. J Nat Prod 62:1427–1429
16. Tafur S, Nelson JD, DeLong DC et al (1976) Antiviral components of *Ophiorrhiza mungos*. Isolation of camptothecin and 10-methoxycamptothecin. Lloydia 39:261–262
17. Arbain D, Putra DP, Sargent MV (1993) The alkaloids of *Ophiorrhiza filistipula*. Aust J Chem 46:977–985
18. Aimi N, Nishimura M, Miwa A et al (1989) Pumiloside and deoxypumiloside; plausible intermediates of camptothecin biosynthesis. Tetrahedron Lett 30:4991–4994
19. Kitajima M, Fujii N, Yoshino F et al (2005) Camptothecin and two new monoterpene glucosides from *Ophiorrhiza liukiuensis*. Chem Pharm Bull 53:1355–1358
20. Darwin SP (1976) The pacific species of Ophiorrhiza L. (Rubiaceae). Lyonia 1:47–102
21. Arbain D, Dachriyanus, Firmansyah et al (1998) Unusual indole alkaloids from *Ophiorrhiza blumeana* Korth. J Chem Soc Perkin Trans 1:2537–2540
22. Aimi N, Hoshino H, Nishimura M et al (1990) Chaboside, first natural glycocamptothecin found from *Ophiorrhiza pumila*. Tetrahedron Lett 31:5169–5172
23. Aimi N, Tsuyuki T, Murakami H et al (1985) Structures of ophiorines A and B; novel type gluco indole alkaloids isolated from *Ophiorrhiza* spp. Tetrahedron Lett 26:5299–5302
24. Aimi N, Murakami H, Tsuyuki T et al (1986) Hydrolytic degradation of β-carboline-type monoterpenoid glucoindole alkaloids: a possible mechanism for harman formation in *Ophiorrhiza* and related Rubiaceous plants. Chem Pharm Bull 34:3064–3066
25. Hayata B (1912) *Ophiorrhiza* Linn. In: Icones of the plants of Formosa, and materials for a flora of the island, based on a study of the collections of the botanical survey of the government of Formosa, fasc. 2. Bureau of Productive Industries, Taihoku, pp 85–92
26. Makino T (1906) Observations on the flora of Japan. Bot Mag 20:1–156
27. van Hengel AJ, Buitelaar RM, Wichers HJ (1994) VII *Camptotheca acuminata* Decne: in vitro cultures and the production of camptothecin. In: Bajaj YPS (ed) Medicinal and aromatic plants VII. Biotechnology in agriculture and forestry, Vol. 28. Springer, Berlin, pp 98–112

28. Sakato K, Tanaka H, Mukai N et al (1974) Isolation and identification of camptothecin from cells of *Camptotheca acuminata* suspension cultures. Agric Biol Chem 38:217–218
29. Wiedenfeld H, Furmanowa M, Roeder E et al (1997) Camptothecin and 10-hydroxycamptothecin in callus and plantlets of *Camptotheca acuminata*. Plant Cell Tissue Organ Cult 49:213–218
30. Roja G, Heble MR (1994) The quinoline alkaloids camptothecin and 9-methoxycamptothecin from tissue cultures and mature trees of *Nothapodytes foetida*. Phytochemistry 36:65–66
31. Ciddi V, Shuler ML (2000) Camptothecine from callus cultures of *Nothapodytes foetida*. Biotechnol Lett 22:129–132
32. Fulzele DP, Satdive RK, Pol BB (2001) Growth and production of camptothecin by cell suspension cultures of *Nothapodytes foetida*. Planta Med 67:150–152
33. Kitajima M, Fisher U, Nakamura M et al (1998) Anthraquinones from *Ophiorrhiza pumila* tissue and cell cultures. Phytochemistry 48:107–111
34. Yamazaki Y, Urano A, Sudo H et al (2003) Metabolite profiling of alkaloids and strictosidine synthase activity in camptothecin producing plants. Phytochemistry 62:461–470
35. De Luca V, Saint-Pierre B (2000) The cell and developmental biology of alkaloid biosynthesis. Trends Plant Sci 5:168–173
36. Kitajima M, Nakamura M, Takayama H et al (1997) Constituents of regenerated plants of *Ophiorrhiza pumila*; formation of a new glycocamptothecin and predominant formation of (3*R*)-deoxypumiloside over (3*S*)-congener. Tetrahedron Lett 38:8997–9000
37. Saito K, Sudo H, Yamazaki M et al (2001) Feasible production of camptothecin by hairy root culture of *Ophiorrhiza pumila*. Plant Cell Rep 20:267–271
38. Asano T, Watase I, Sudo H et al (2004) Camptothecin production by *in vitro* cultures of *Ophiorrhiza liukiuensis* and *O. kuroiwai*. Plant Biotech 21:275–281
39. Watase I, Sudo H, Yamazaki M et al (2004) Regeneration of transformed *Ophiorrhiza pumila* plants producing camptothecin. Plant Biotech 21:337–342
40. Stöckigt J, Zenk MH (1977) Strictosidine (isovincoside): the key intermediate in the biosynthesis of monoterpenoid indole alkaloids. J Chem Soc Chem Commun 18:646–648
41. Kutchan TM, Hampp N, Lottspeich F et al (1988) The cDNA clone for strictosidine synthase from *Rauvolfia serpentina*. DNA sequence determination and expression in *Escherichia coli*. FEBS Lett 237:40–44
42. Stöckigt J, Ruppert M (1999) Strictosidine—the biosynthetic key to monoterpenoid indole alkaloids. In: Barton SD, Nakanishi K, Meth-Cohn O (eds) Comprehensive natural products chemistry, Vol. 4. Pergamon, Oxford, pp 109–138
43. Hutchinson CR, Heckendorf AH, Straughn JL et al (1979) Biosynthesis of camptothecin. 3. Definition of strictosamide as the penultimate biosynthetic precursor assisted by ^{13}C and ^{2}H NMR spectroscopy. J Amer Chem Soc 101:3358–3369
44. Carte BK, DeBrosse C, Eggleston D et al (1990) Isolation and characterization of a presumed biosynthetic precursor of camptothecin from extracts of *Camptotheca acuminata*. Tetrahedron 46:2747–2760
45. Noé W, Mollenschott C, Berlin J (1984) Tryptophan decarboxylase from *Catharanthus roseus* cell suspension cultures: purification, molecular and kinetic data of the homogenous protein. Plant Mol Biol 3:281–288
46. Yamazaki Y, Sudo H, Yamazaki M et al (2003) Camptothecin biosynthetic genes in hairy roots of *Ophiorrhiza pumila*: cloning, characterization and differential expression in tissues and by stress compounds. Plant Cell Physiol 44:395–403
47. McKnight TD, Roessner CA, Devagupta R et al (1990) Nucleotide sequence of a cDNA encoding the vacuolar protein strictosidine synthase from *Catharanthus roseus*. Nucl Acids Res 18:4939
48. De Luca V, Marineau C, Brisson N (1989) Molecular cloning and analysis of a cDNA encoding a plant tryptophan decarboxylase: comparison with animal DOPA decarboxylases. Proc Natl Acad Sci U S A 86:2582–2586

49. López-Meyer M, Nessler CL (1997) Tryptophan decarboxylase is encoded by two autonomously regulated genes in *Camptotheca acuminata* which are differentially expressed during development and stress. Plant J 11:1167–1175
50. Meijer AH, Lopes Cardoso MIL, Voskuilen JT et al (1993) Isolation and characterization of a cDNA clone from *Catharanthus roseus* encoding NADPH:cytochrome P450 reductase, an enzyme essential for reactions catalysed by cytochrome P450 mono-oxygenases in plants. Plant J 4:47–60
51. Cane DE (1999) Isoprenoid biosynthesis overview. In: Cane DE (ed) Comprehensive natural product chemistry, vol. 2. Elsevier, Amsterdam, pp 1–13
52. Rohmer M (1999) A mevalonate-independent route to isopentenyl diphosphate. In: Cane DE (ed) Comprehensive natural product chemistry, vol. 2. Elsevier, Amsterdam, pp 45–67
53. Rodríguez-Concepción M, Boronat A (2002) Elucidation of the methylerythritol phosphate pathway for isoprenoid biosynthesis in bacteria and plastids. A metabolic milestone achieved through genomics. Plant Physiol 130: 1079–1089
54. Kuzuyama T, Seto H (2003) Diversity of the biosynthesis of isoprene units. Nat Prod Rep 20: 171–183
55. Yamazaki Y, Kitajima M, Arita M et al (2004) Biosynthesis of camptothecin. In silico and in vivo tracer study from [1-^{13}C]glucose. Plant Physiol 134:161–170
56. Arita M (2003) *In silico* atomic tracing by substrate-product relationships in *Escherichia coli* intermediary metabolism. Genome Res 13:2455–2466
57. Park SU, Yu M, Facchini PJ (2002) Antisense RNA-mediated suppression of benzophenanthridine alkaloid biosynthesis in transgenic cell cultures of California poppy. Plant Physiol 128:696–706
58. Frick S, Chitty JA, Kramell R et al (2004) Transformation of opium poppy (*Papaver somniferum* L.) with antisense berberine bridge enzyme gene (*anti-bbe*) via somatic embryogenesis results in an altered ratio of alkaloids in latex but not in roots. Transgenic Res 13:607–613
59. Allen RS, Millgate AG, Chitty JA et al (2004) RNAi-mediated replacement of morphine with the nonnarcotic alkaloid reticuline in opium poppy. Nat Biotechnol 22:1559–1566
60. Asano T, Kobayashi K, Kashihara E et al (2013) Suppression of camptothecin biosynthetic genes results in metabolic modification of secondary products in hairy roots of *Ophiorrhiza pumila*. Phytochemistry 91:128–139

Chapter 4
Plant Defense Activators: Application and Prospects in Cereal Crops

Mitchell L. Wise

Abstract Cereal grains are one of the primary sources of food products in the world. Increased productivity in crop yield, particularly for cereal crops, is absolutely essential for future food security, but is impeded by disease, with annual estimates ranging from 10 to 30% crop loss due to disease alone. There have been remarkable advances in understanding pest and disease resistance in plants in the past three decades, with the application of chemical plant defense activators (PDAs) being of particular interest. The advances in recent years in understanding the molecular basis for systemic acquired resistance (SAR), induced systemic resistance (ISR), priming, and next-generation immunity portend a wider role for PDAs. These agrochemicals are gaining some acceptance in Europe where there is a strong interest in curtailing the use of more traditional fungicides and pesticides. Much work, however, is needed to understand the effects of nutrition, dose rates, timing of application, and genotypic effects in the application of PDAs. This review addresses the current understanding of plant immunity, particularly with respect to cereal crops and the potential for PDAs to enhance the potential yield and nutritional quality of cereal crops.

4.1 Introduction

As the world population increases from the current estimate of 7 billion to a projected 9 billion by 2050 [1] and as greater demand on land usage for activities other than agriculture increases, food supply will become an issue of even greater importance. Cereal grains are one of the primary sources of food products in the world today [2]; this is unlikely to change in the foreseeable future. Thus, increased productivity in crop yield, particularly for cereal crops, is absolutely essential for future

M. L. Wise (✉)
Cereal Crops Research Unit, USDA, ARS, Madison, WI, USA
e-mail: mlwise@wisc.edu

D. R. Gang (ed.), *50 Years of Phytochemistry Research*,
Recent Advances in Phytochemistry 43, DOI 10.1007/978-3-319-00581-2_4,

food security. One of the major impediments to crop yield is disease, with annual estimates ranging from 10 to 30 % crop loss due to disease alone [1]. In addition to yield loss, pathogens can also contaminate food crops with toxins, rendering them useless, when detected, and dangerous for human consumption when left undetected. Disease loss, however, can be mitigated by informed agricultural practices.

There have been remarkable advances in understanding pest and disease resistance in plants in the past three decades. Much of this research has been conducted on dicots, *Arabidopsis* and tobacco being the two principal model organisms. After the physical barrier of the leaf cuticle or outer cell wall of other tissues, the fundamental disease resistance mechanism in plants is its basal resistance. Our understanding of this phenomenon has undergone dramatic changes in the last two decades, leading to a much keener understanding of the molecular events and signaling mechanisms involved in plant "immunity." As a result, improved methods to elicit this response have come to fruition. One of these methods is the application of chemical plant defense activators (PDAs). This review addresses the current understanding of plant immunity, particularly with respect to cereal crops and the potential for PDAs to enhance the potential yield and nutritional quality of cereal crops.

4.2 Pathogen Recognition

The prevailing model in plant disease resistance for most of the past century has been the gene-for-gene theory, based on the pioneering genetic studies of Harold H. Flor [3]. This theory posits that plants recognize microbial pathogens by their avirulence factors and combat them through expression of resistance genes, termed "R" genes. More recently, a "zigzag" model (Fig. 4.1) of pathogen resistance has emerged [4]. This paradigm rationalizes that plants recognize a pathogen invader first by interaction of pathogen-associated molecular patterns (PAMPs) (sometimes called microbe-associated molecular patterns (MAMPs)). The presence of these molecules at the plant cell membrane suggests attack by a potential pathogen. Bacterial flagellin, certain lipopolysaccharides, and chitin (polymeric *N*-acetyl glucosamine, a component of many fungal species cell walls and, coincidentally, crustacean shells) are well-documented examples of MAMPs. Plant cell surface receptors called pattern recognition receptors (PRR) interact with MAMPs to trigger the initial stage of plant defense, termed PAMP-triggered immunity (PTI). PRRs are transmembrane proteins consisting of an extracellular leucine-rich repeat motif and, typically, an intracellular protein kinase. Only a few PRRs have been characterized for their specific binding mechanisms. Perhaps the most thoroughly studied examples are the *Arabidopsis* receptor for bacterial flagellin peptide [5] and the receptor for bacterial elongation factor-Tu [6, 7]. For the most part, these PRRs recognize highly conserved pathogen-derived molecules. MAMPs appear to be essential to the survivability and pathogenicity of the offending organism, thus not being readily adaptable to mutation. Equally important is the ability of the plant to discern these MAMPs from beneficial microbes and even its own molecular patterns, thus

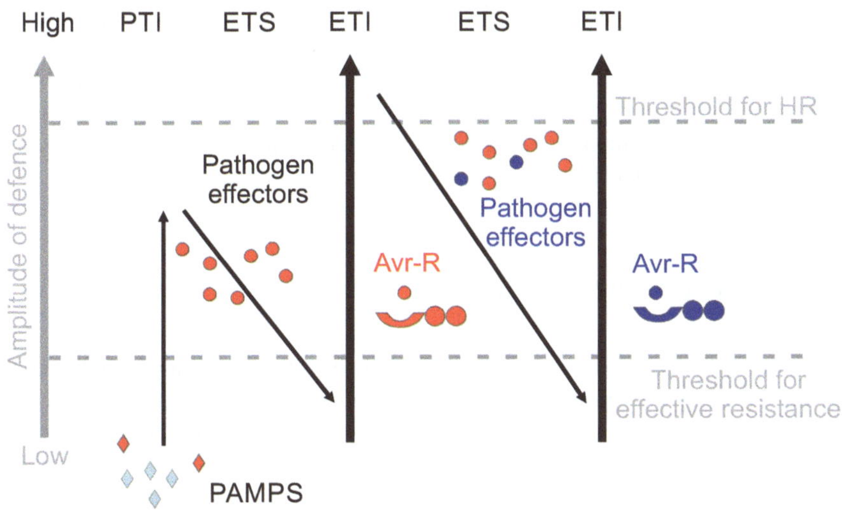

Fig. 4.1 The "zigzag" model of pathogen resistance

avoiding complications arising from autoimmunity [8]. Plants do, however, have the ability to detect self-antigens in the form of damage-associated molecular patterns (DAMPs) typically resulting from herbivore damage or microbial pathogen-mediated lytic enzymes (see [9] for an excellent review of the biochemistry of plant PRRs).

Upon activation, the PRRs initiate a plethora of defense mechanisms, including callose deposition in the cell wall, increased ion flux, particularly the influx of H^+ and Ca^{++} and efflux of K^+ ions, activation of plant mitogen-activated protein kinases (MAPKs) with concurrent phosphorylation of numerous signaling proteins, the generation of reactive oxygen species, biosynthesis of pathogenesis-related (PR) proteins,and the production of phytoalexins.

To overcome the basal immunity elicited through PTI, pathogenic microbes have evolved another strategy to circumvent the signaling mechanisms triggered by their MAMPs. This second phase of the zigzag model involves the delivery of effectors into the plant cell. Effectors are, to a large extent, what was previously termed virulence (or avirulence) factors in Flor's gene-for-gene theory of disease resistance. In essence, effectors are an array of pathogen-derived metabolites and proteins that interfere with host defense mechanisms. Pathogenic bacteria typically inject their effectors into the host cytoplasm through a type-three secretory system [10]. Introduction of effectors by other eukaryote pathogens is not as well understood; however, evidence points to RxLR-EER protein motifs, similar to those employed by *Plasmodium* (malarial) parasites in mammals [11]. This motif binds to phosphatidylinositol-3-phosphate moieties in the cell membrane [12], whereupon they are translocated into the cytoplasm. This likely represents a mechanism for both fungal and oomycete effectors [12, 13]. Effectors work through a wide range of mechanisms. Small molecule effectors like coronatine, a non-host-specific phytotoxin produced by pathovars of the bacterial pathogen *Pseudomonas syringae*, appear to mimic the action of jasmonic acid (JA) [4, 14, 15], thus suppressing the effect of salicylic acid (SA) [16, 17]. Effectors also abrogate the defense response through other, largely unknown, mechanisms. Many clearly inhibit the host defense signaling pathways [18, 19].

4.3 Salicylic Acid Signaling

Recognition of pathogen or herbivore invasion results in what is now considered two separate signaling pathways that elicit an enhanced resistance response at locations distal to the site of infection. The first of these pathways was described by Ross in 1961 in which he demonstrated that tobacco leaves inoculated with tobacco mosaic virus (TMV) produced a lasting immunity in other portions of the plant against this virus as well as other viral pathogens. He termed this response systemic acquired resistance (SAR) [20]. This phenomenon has subsequently been demonstrated in numerous host–pathogen relationships and appears to be a characteristic of most, if not all, terrestrial plants [21]. The nature of the mobile signaling molecule(s) has been a subject of intense research and not inconsiderable controversy since SAR was initially proposed. Acetylsalicylic acid along with salicylic and benzoic acid were demonstrated to induce resistance against TMV in tobacco plants in 1979 [22]. Subsequently, SA and its methyl ester have been presented as the likely candidate as the mobile messenger in SAR [23, 24] and, in fact, a chemical analog for SA, 2,6-dichloroisonicotinic acid (INA), can replace SA in eliciting SAR in *Arabidopsis* and tobacco plants deficient in SA biosynthesis [25]. At present, SA is generally accepted as a key molecular component in SAR signaling [26]. Indeed, a chemical rationale for SA activation of defense responses has been demonstrated. Xinnian Dong and her colleagues' pioneering work has revealed important relationships between SA accumulation, PR protein expression, and activation of the nonexpressor of the PR (NPR1) protein [27]. NPR1 is so named because *Arabidopsis* mutants deficient in this protein do not respond to the normal signaling mechanisms for PR gene expression as well as a number of other pathogen defense-related genes [28]. This phenotype has also been called NIM1 and SAI1 [29]. NPR1 is now recognized as an essential regulator of plant defense mechanisms that normally resides in the cytosol of plant cells as a multimeric complex. This complex is maintained through redox-sensitive disulfide bonds [30, 31] that, under reducing conditions resulting from, for example, high concentrations of SA, partially disassociate into monomers [32]. The monomeric form of NPR1 is subsequently transported to the cell nucleus where it serves as a gene transcription coactivator [31]. Interestingly (and almost paradoxically), Spoel et al. [31] have recently shown that the full expression of SAR requires that NPR1 be imported into the nucleus and then be ubiquitinylated and degraded by nuclear proteasomes. This process is hypothesized to facilitate clearance of the NPR1-transcription factor (TF)/polymerase complex to allow fresh NPR1-TF

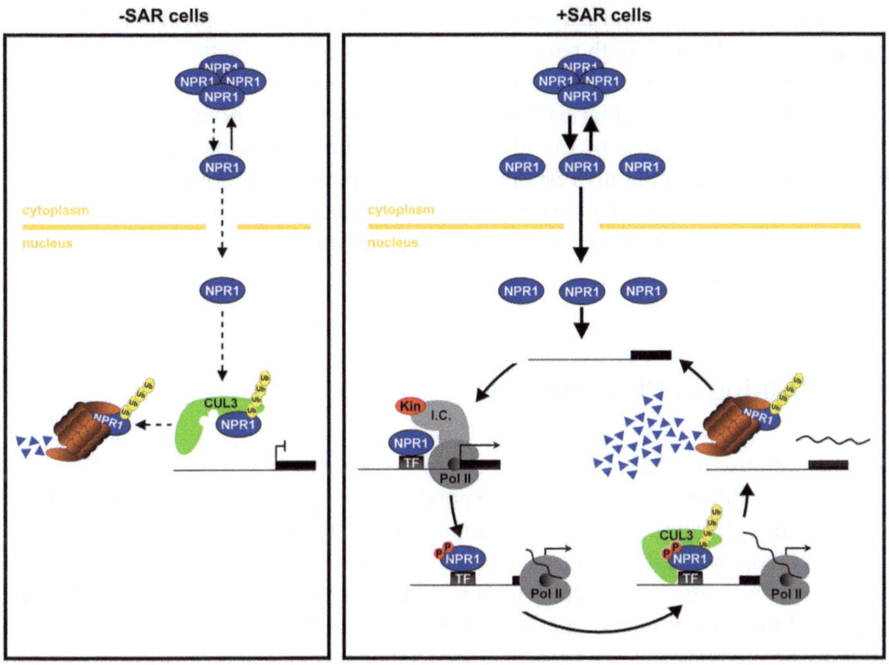

Fig. 4.2 The role and activation of NPR1 in SAR+ and SAR− cells

access to the promoter region of affected genes in order to maintain transcription activity. Ubiquitinylation appears to rely on phosphorylation of specific serine residues in the NPR1 protein (Fig. 4.2).

Additional mobile signals are being recognized as mediators of SAR. Azelaic acid [33] and glycerol-3-phosphate [34] are two small molecules recently associated with SAR. Both of these also require expression of DIR1 (defective in induced resistance) protein [35]. There appears to be additional, as yet undetermined, mobile signals in the phloem exudates from *Arabidopsis* associated with azelaic acid signaling [34].

Airborne signals can also contribute to plant defense against pests and pathogens [36]. Notably, JA and related jasmonates are known for their role in systemic responses to wounding by insect herbivores [37]. The methyl ester of JA and certain other jasmonates are also quite volatile and have been implicated in airborne signaling to nearby plants [36]. Other volatile metabolites, such as short-chain oxylipins and terpenoids, also function in plant-to-plant defense signaling as well as complex tritrophic interactions involving plant pests' predators [38].

The second of the two major defense pathways is termed induced systemic resistance (ISR)and is a systemic immune response elicited primarily by plant growth-stimulating rhizobacteria [39] and certain rhizosphere-associated fungi [40]. Similar to SAR in many respects, ISR, however, does not require SA; it is more dependent on JA and ethylene signaling. ISR does not typically result in PR accumulation, although both pathways depend on NPR1 activation [41, 42].

Thus, plants have evolved intricate systems to defend themselves against herbivorous pests and microbial pathogens. These defenses involve signaling mechanisms to alert distal parts of the plant, or even other plants, of impending attack. Because plants must not only respond to invading pest and pathogens but also avoid autoimmune responses, or unnecessarily responding to a plethora of non-harmful (or even beneficial) microorganisms, multiple and complex signaling mechanisms should be expected.

4.4 Priming in Plant Defense

Another defense mechanism related to induced resistance is "priming." This is defined as a condition in which plants respond faster or more strongly than unprimed plants in the activation of defense responses when subsequently challenged by microbial pathogens, herbivorous insects, or abiotic stresses [43]. This phenomenon occurs after an initial encounter with a pathogen or chemical elicitor but without a display of the typical phenotypes of induced resistance such as upregulation of PR proteins or phytoalexins. Because the molecular mechanisms of priming are just now being determined (phenotypic analysis of priming has relied on tedious post-challenge defense responses [44]), this phenomenon has likely been overlooked in many SAR studies. Uwe Conrath and coworkers demonstrated that cultured parsley cells, pretreated with low concentrations of INA, responded to subsequent elicitation with a known fungal MAMP with dramatically higher levels of coumarins and phenylalanine ammonia lyase (PAL) activity, SAR biomarkers [45]. Similar results were observed using benzo (1,2,3) thiadiazole (BTH), another synthetic SA analog [46]. It is important to note that this "priming" was dependent on the dose rate of INA or BTH, with relatively low doses resulting in priming, and higher doses resulting in elicitation of SAR [43]. Thus, priming appears to potentiate the plant for subsequent pathogen or pest challenge. This phenotypic difference is important because direct activation of plant defense mechanisms appears to extract a fitness cost that may, for example, reduce seed set [47] (discussed later).

Over the past 3–5 years, inroads have been achieved in determining the molecular mechanisms responsible for priming. *Arabidopsis* plants treated with BTH under priming conditions were shown to upregulate the levels of two mitogen-activated protein kinases, MPK3 and MPK6. These are cytosolic elements that transmit and amplify extracellular stimuli from external receptors into intracellular responses through a series of protein phosphorylation reactions. This study convincingly demonstrated that *Arabidopsis* plants primed with BTH responded far more strongly to biotic and abiotic stress in producing SAR biomarkers such as PR proteins and PAL. The primed plants also proved significantly more refractive to bacterial infection. Use of mpk3- and mpk6-deficient mutants conclusively demonstrated the involvement of these two enzymes in priming [48, 49].

Priming was also shown to modify chromatin associated with the promoter regions of certain WRKY genes in *Arabidopsis*. Methylation and acetylation of histones are instrumental in gene regulation [50, 51] and can result in long-term activation (or suppression) of the associated gene [52, 53]. WRKYs are transcription factors closely allied with many defense-related genes in plants [54–56]. Using chromatin immunoprecipitation, Jaskiewicz et al. [57] recently demonstrated that histones bound to certain WRKY promoter regions are methylated and or acetylated in response to priming by either BTH treatment or exposure to a pathogen. Thus, chromatin modification appears to be involved in priming.

Epigenetic control of plant immunity can also be manifested through DNA methylation, which, in turn, can even result in enhanced pathogen resistance in progeny plants. This was recently demonstrated in *Arabidopsis* which the investigators termed "next-generation SAR" [58]. These investigators found that when *Arabidopsis* plants were repeatedly inoculated with the bacterial pathogen *Pseudomonas syringae* pv. *tomato* DC3000 (*Pst*DC3000) their first-and even second-generation progeny were more resistant to infection with the oomycete pathogen *Hyaloperonospora arabidopsidis*. Use of *Arabidopsis* mutant lines further demonstrated that this transgenerational resistance was dependent on functional NPR1 activation. Additionally, triple mutants deficient in DNA methylation were constitutively more resistant to *Pst*DC3000 infection, leading these investigators to suggest that hypomethylation plays a role in next-generation resistance [58]. Hypomethylation has been observed previously in *Arabidopsis* in response to *P. syringae* attack [59] and was specifically associated with defense-related gene expression in tobacco infected with TMV [60]. Similar transgenerational enhancement of SAR response, as determined by PR-1 biosynthesis and resistance to bacterial and oomycete challenge, was observed in *Arabidopsis* treated with β-aminobutyric acid (BABA), another of the chemical SA mimics [61]. Although in this case the next-generation priming only appeared to last through one generation with respect to PR-1 expression, second-generation BABA-treated plants did retain some resistance to pathogen challenge.

4.5 Commercial Plant Defense Activators

Advances in our understanding of the plant immune response have resulted in, and in some cases resulted from, the development of numerous synthetic compounds that appear to mimic the effect of SA in stimulating plant immunity. Some of these are currently marketed for commercial field application. Interestingly, the first of these commercial products was probenazole (3-allyoxy-1,2-benzisothiazole-1,1-dioxide) marketed under the trade name Oryzemate®, which proved effective in reducing rice blast infection. Developed in the mid-1970s, it was almost 20 years later that this compound was recognized as being effective against certain bacterial pathogens as well. Only due to advances in understanding SAR and ISR was the mechanism of resistance revealed [62, 63]. A systematic investigation of synthetic chemical inducers of SAR by investigators at Ciba-Geigy in the early 1990s resulted in the discovery of INA and derivatives [64]. Although highly effective in some plants, INA was not well tolerated by others and was never developed commercially [65]. Subsequently, benzothiadiazole (BTH) derivatives were found to be just as effective in eliciting SAR but better tolerated in a broader range of crops. Formulations of these compounds have been marketed under the trade names BION® in Europe and Actigard® in the USA [62]. A more recent addition among the thiadiazole derivatives is 3′-chloro-4,4′-dimethyl-1,2,3-thiadiazole-5-carboxanilide, common name tiadinil, marketed by Nihon Nohyaku Co, Led with the trade name V-GET® for use against rice blast [66]. A bacterial pathogen effector protein, harpin, has been used to successfully combat blue mold in apples [67]. Fragments of the harpin protein also stimulated plant growth and disease resistance in rice in field trials [68]. A commercial formulation of harpin was originally marketed by Enden Bioscience as Messenger®; it has recently been replaced by Employ® from Plant Health Products. A de-acetylated form of chitin (chitosan) has been produced as a commercial product named Elexa® in a 4 % chitosan formulation. Greenhouse as well as field trials of Elexa® on pearl millet (*Pennisetum glaucum*) administered either as a seed treatment or as a foliar spray or in combination showed it to be highly effective at inducing resistance to downy mildew disease caused by *Sclerospora graminicola* [69]. Although chitin and chitosan are well-known elicitors of SAR, chitosan also shows some fungicidal activity [70]. BABA is a nonprotein amino acid that also induces systemic resistance in plants. The amino acid has been used extensively in experimental systems to induce SAR as well as priming. Its commercial application has not been realized.

4.6 Induced Resistance in Cereal Crops

The bulk of the research on induced resistance has been performed on dicots, particularly the model plants *Arabidopsis* and tobacco. Nevertheless, monocots are capable of generating SAR and ISR and thus likely possess all the requisite signaling and defense activators found in dicots [71–74]. Indeed, BTH was originally developed to protect wheat from a variety of fungal pathogens. It proved effective in both growth chamber and field experiments [65]. These results were somewhat telling because although an important role of SA in dicot innate immunity is well established [26], an analogous role in monocots was not obvious at the time of these experiments. Rice (*Oryza sativa*) had been shown to constitutively produce dramatically higher levels of SA than healthy tobacco plants and these levels did not appear to be affected by infection with avirulent or virulent pathogens [75]. Thus, early investigations into the mechanism of SAR in cereals cast doubt on the role of SA. In addition, while INA and BTH proved effective in eliciting SAR-like responses in cereals [65, 76], the suite of defense-related gene expression appears to differ from dicots. Specifically, PR-1-related protein expression was not observed in wheat treated with INA or BTH [77]. Thus, some investigators did not consider PR gene expression a particularly reliable biomarker for SAR in monocots, although other "chemically inducible" genes have been described in wheat, rice, and barley [78]. Nevertheless, as more research on defense mechanisms in monocots is published, the similarity to dicots becomes more apparent [72–74, 79]. Rice, for example, has now been shown to have an SA/NPR1-mediated defense network similar to *Arabidopsis* [80]. Thus, employment of PDAs on cereal crops appears to be perfectly feasible, including those that mimic SA, at least in terms of eliciting a defense response. The method of PDA application on cereals, however, can be a critical factor. Initial efforts to induce resistance to *Fusarium* head blight (FHB) in wheat, through the application of BTH as a foliar spray, proved ineffective [81]. More recently, both SA and BTH proved highly effective in protecting wheat against the same pathogen (*Fusarium graminearum*) in greenhouse trials when applied as a root soak [82].

Another aspect of PDA application that warrants consideration is their ability to elicit phytoalexin biosynthesis. Oat (*Avena sativa*) plants treated with SA or BTH were recently shown to dramatically increase their production of avenanthramides. Avenanthramides are phenolic alkaloid compounds produced, among food crops, exclusively by oats. They are known to function as phytoalexins in the vegetative tissue in response to crown rust (*Puccinia coronata*) infection [83, 84]. These metabolites are also potent antioxidants that, in laboratory trials, show potential as phytonutrients [85]. Unfortunately, the levels of avenanthramides found in the grain are highly variable and subject to strong environmental influence [86], and there appears to be an association between crown rust infection and avenanthramide content in the mature grain under field conditions [87]. The means of enhancing the levels of grain avenanthramides is of interest. BTH treatment in the form of Actigard® was recently shown to strongly induce avenanthramide biosynthesis in vegetative tissue of oat seedlings when administered as a root soak in greenhouse experiments [88]. RNA hybridization (Northern) analysis also showed elicitation of an RNA message hybridizing with a barley PR-1 probe, suggesting that the avenanthramide production was part of a SAR response. Moreover, when mature plants were treated with BTH just prior to heading, certain cultivars showed a statistically significant increase in avenanthramide content in the filling grain [89]. Indeed, all of the treated oat cultivars were higher in grain avenanthramide content than the untreated controls. However, since oat constitutively produces avenanthramides in their grain, and these levels show high variability, it can be difficult to establish a statistically significant difference. Nevertheless, these findings portend the utility of PDAs to upregulate the biosynthesis of avenanthramides in oat and, by extension, may be of use in other crops whose phytonutrient content is augmented by natural phytoalexins. Harpins, for example, were shown, in field trails, to increase yield and catechol levels in green tea [90].

4.7 Fitness Costs

The evolutionary rationale for induced resistance holds that plants cannot afford to biosynthesize the pest and pathogen defense metabolites on a constitutive basis because of either detrimental allocation of nutrient resources, production of autotoxic metabolites, or negative effects on beneficial microorganisms [67, 91]. Thus, it is better to resort to these biosynthetic pathways only when they are in dire need. The application of PDAs circumvents this "just-in-time" approach evolved in plants. In most cases, field studies have focused on the efficacy of PDAs to reduce disease pressure. One study specifically aimed at determining the "fitness" cost of BTH application was conducted with spring wheat (*Triticum aestivum* cv 'Hanno') by treating them with BION® under a variety of treatment regimens and growing them either hydroponically with carefully controlled nutrient conditions or in a field environment with added fertilizer and fungicide treatment to ensure no disease pressure [47]. The investigators reported a statistically significant reduction in seed production and growth rates in the BTH-treated plants relative to uninduced control plants when nitrogen availability was limited. They suggested that this likely represented an allocation cost to chemically induced resistance in the absence of disease pressure. It should be noted, however, that in those plants treated late in their growth cycle or provided adequate nitrogen no significant differences in the measured parameters were observed between the BION®-treated and the untreated controls [47]. Several studies comparing the yield of various cereals treated with BTH versus standard pesticides provide little evidence for increased yield from BTH treatment (summarized in [42]). Treatment of plants with PDAs mimicking the SA elicitation pathway can also prove antagonistic to JA signaling. A study conducted with two barley (*Hordeum vulgare*) cultivars, 'Celler' and 'Optic', treated with a combination of BION®, BABA, and *cis*-jasmone under field conditions resulted in a marked decrease in infection levels by the biotrophic pathogen *Blumeria graminis* and the hemibiotroph *Rhynchosporium secalis*, etiological agents of powdery mildew and leaf scald, respectively. Infection by *Ramularia collo-cygni* (*Ramularia* leaf spot), another hemibiotroph, however, was significantly higher in the treated plants. Analysis of PR-1 and lipoxygenase (LOX2, an enzyme involved in JA biosynthesis) showed that elicitor treatment upregulated PR-1 expression, whereas LOX2 expression was downregulated. The combination of PDAs was used to specifically target *R. secalis*, making the interpretation of the results somewhat complicated. Note, also, that the grain yield from the elicitor-treated cultivars was slightly higher than the controls in both years of this study, although the authors speculated that the mixed result in protection might be due to suppression of the JA signaling, this pathway possibly being more important in defense against *R. collo-cygni* infection [92]. BTH-treated tomato plants have shown enhanced resistance to *Pseudomonas syringae* pv. *tomato* but compromised resistance to herbivore attack by *Spodoptera exigua* and, conversely, treatment with JA enhanced herbivore resistance at the expense of bacterial infection [93]. Thus, antagonism between SA and JA signaling may result in tradeoffs in the protective effects of PDAs in some situations [67].

In contrast to PDA-induced resistance, priming seems to have minimal allocation costs. Laboratory trials using *Arabidopsis* treatment with BABA in a range of concentrations resulted in induced direct defense response (as determined by PR-1 biosynthesis) at higher concentrations and priming of the plants at lower treatment levels. The primed plants were only slightly less resistant to subsequent challenge with either the bacterial pathogen *P. syringae* or the fungal pathogen *Hyaloperonospora parasitica* but did not demonstrably increase PR-1 levels prior to the challenge [94]. All the PDA-treated plants, including control plants fully induced with BTH, were significantly more resistant to infection than the mock-induced controls. Moreover, fitness costs, evaluated in terms of seed set and relative growth rates, were only marginally affected in the BABA-primed plants; plants induced to direct defense levels demonstrated significantly lower fitness levels.

Saccharin is a metabolic by-product of probenazole (Oryzamate®) that can induce priming in barley [95]. In a study on fitness costs of saccharin-induced priming in barley (*H. vulgare* cv 'Celler'), Walters et al. found that, in greenhouse experiments, priming did not incur significant costs under low disease pressure by the hemibiotrophic fungal pathogen *R. secalis* and that it provided significant benefits under high disease pressure [96]. Use of saccharin-primed barley, in a field environment subject to low disease pressure from three fungal pathogens, *R. secalis, B. graminis, and R collo-cygni*, similarly showed little or no fitness costs, although the application of commercial fungicide resulted in significantly higher grain yield in adjacent plots [96].

4.8 Plant Defense Activators, Prospects for Cereal Crops

The commercial application of PDAs in cereal crops has been met with limited enthusiasm [67, 97]. Probenazole, as Oryzemate®, has been used for over three decades and remains one of the major fungicides used for seedling box treatment of rice [98]. It is noteworthy that no resistance to this product has yet developed [62]. Other commercial formulations such as BION®/Actigard®, Messenger®, and Elexa® have found use primarily on vegetable crops [62]. Some of the major drawbacks to PDAs are their unreliability under field conditions [99]. Abiotic and biotic factors affecting induced resistance in commercial/field application are still poorly understood. Numerous parameters must be further investigated. For example, the timing of application may be critical [99, 100]. Recent evidence has even shown that plant immune responses, particularly those associated with SA, are sensitive to light intensity and circadian rhythms [101]. Genotypic effects have also received little attention. Oat cultivars, for example, show dramatically different responses in both the magnitude and kinetics of avenanthramide biosynthesis in response to BTH and INA treatment [89]. When seven cultivars of spring barley were assessed for the efficacy of induced resistance to *R. secalis* and powdery mildew (*B. graminis*) induced by treatment with a suite of PDAs (BION®, BABA and *cis*-jasmone), significant differences were observed between cultivars [102]. A few additional examples are outlined by Walters and Fountaine [67].

PDAs are not curative; they must be administered prior to pathogen or pest invasion; thus, any fitness costs incurred may, in fact, be as detrimental to crop yield as the pathogen itself. However, an integrated crop management approach where PDAs are used in concert with more traditional fungicides/pesticides or other agents such as biocontrol or plant growth promoting rhizobacteria might have some merit, particularly in reducing fungicide treatment levels [103]. Certainly more research on fertilizer augmentation to ameliorate allocation costs is warranted. PDAs can elicit volatile signals yielding protective effects against bacterial infection on neighboring plants as was recently demonstrated in lima bean (*Phaseolus lunatus*) [104]. This suggests the possibility of treating border rows, for example, to enhance resistance in the larger field. Maize is well known to produce volatile organic compounds (VOCs) in response to herbivore attack [105, 106]. There is evidence that these airborne signals can prime neighboring plants to respond to subsequent herbivore attack [107, 108]. Cereal crops also release VOCs in response to fungal infection [109], although the chemical ecology of these emissions is poorly understood. Indeed, in spite of an extensive literature on the generation of VOCs in response to herbivore attack, where cereal crops such as corn and rice are well represented [110], there is little research on the chemical ecology of VOC emission resulting from pathogen infection, especially in cereal crops.

The advances in recent years in understanding the molecular basis for SAR, ISR, priming, and next-generation immunity portend a wider role for PDAs. These agrochemicals are gaining some acceptance in Europe where there is a strong interest in curtailing the use of more traditional fungicides and pesticides. Much work, however, is needed to understand the effects of nutrition, dose rates, timing of application, and genotypic effects in the application of PDAs. Exploitation of plant immunity can and should be a useful tool in our collective arsenal for combating plant disease.

References

1. Bennett AJ et al (2012) Meeting the demand for crop production: the challenge of yield decline in crops grown in short rotations. Biol Rev Camb Philos Soc 87:52
2. Strange RN, Scott PR (2005) Plant disease: a threat to global food security. Annu Rev Phytopathol 43:83
3. Flor HH (1955) Host-parasite interaction in Flax rust-its genetics and other implications. Phytopathology 45:680
4. Jones JDG, Dangl JL (2006) The plant immune system. Nature 444:323
5. Felix G et al (1999) Plants have a sensitive perception system for the most conserved domain of bacterial flagellin. Plant J 18:265
6. Kunze G et al (2004) The N terminus of bacterial elongation factor Tu elicits innate immunity in arabidopsis plants. Plant Cell 16:3496
7. Zipfel C et al (2006) Perception of the bacterial PAMP EF-Tu by the receptor EFR restricts agrobacterium-mediated transformation. Cell 125:749
8. Spoel SH, Dong X (2012) How do plants achieve immunity? Defense without specialized immune cells. Nat Rev Immunol 12:89
9. Boller T, Felix G (2009) A renaissance of elicitors: perception of microbe-associated molecular patterns and danger signals by pattern-recognition receptors. Annu Rev Plant Biol 60:379
10. He SY (1998) Type III protein secretion systems in plant and animal pathogenic bacteria. Annu Rev Phytopathol 36:363
11. Whisson SC et al (2007) A translocation signal for delivery of oomycete effector proteins into host plant cells. Nature 450:115

12. Kale SD et al (2010) External lipid PI3P mediates entry of eukaryotic pathogen effectors into plant and animal host cells. Cell 142:284
13. Bozkurt TO et al (2012) Oomycetes, effectors, and all that jazz. Curr Opin Plant Biol 15:483
14. Kloek AP (2001) Resistance to *Pseudomonas syringae* conferred by an *Arabidopsis thaliana* coronatine-insensitive (coi1) mutation occurs through two distinct mechanisms. Plant J 26:509
15. Uppalapati SR et al (2007) The phytotoxin coronatine contributes to pathogen fitness and Is required for suppression of salicylic acid accumulation in tomato inoculated with Pseudomonas syringae pv. tomato DC3000. Mol Plant Microbe Interact 20:955
16. Beckers GJM, Spoel SH (2006) Fine-tuning plant defense signaling: salicylate versus jasmonate. Plant Biol 8:1
17. Kunkel BN, Brooks DM (2002) Cross talk between signaling pathways in pathogen defense. Curr Opin Plant Biol 5:325
18. Bent AF, Mackey D (2007) Elicitors, effectors, and R genes: the new paradigm and a lifetime supply of questions. Annu Rev Phytopathol 45:399
19. De Wit PJGM et al (2009) Fungal effector proteins: past, present and future. Mol Plant Pathol 10:735
20. Ross AF (1961) Systemic acquired resistance induced by localized virus infections in plants. Virology 14:340
21. Sticher L et al (1997) Systemic acquired resistance. Annu Rev Phytopathol 35:235
22. White RF (1979) Acetylsalicylic acid (aspirin) induces resistance to tobacco mosaic virus in tobacco. Virology 99:410
23. Gaffney T et al (1993) Requirement of salicylic acid for the induction of systemic acquired resistance. Science 261:754
24. Malamy J et al (1990) Salicylic acid: a likely endogenous signal in the resistance response of tobacco to viral infection. Science 250:1002
25. Delaney TP et al (1994) A central role of salicylic acid in plant disease resistance. Science 266:1247
26. Vlot AC et al (2009) Salicylic acid, a multifaceted hormone to combat disease. Annu Rev Phytopathol 47:177
27. Dong X (2004) NPR1, all things considered. Curr Opin Plant Biol 7:547
28. Cao H et al (1994) Characterization of an Arabidopsis mutant that Is nonresponsive to inducers of systemic acquired resistance. Plant Cell 6:1583
29. Glazebrook J (2001) Genes controlling expression of defense responses in Arabidopsis: 2001 status. Curr Opin Plant Biol 4:301
30. Spoel SH, Loake GJ (2011) Redox-based protein modifications: the missing link in plant immune signalling. Curr Opin Plant Biol 14:358
31. Spoel SH et al (2009) Proteasome-mediated turnover of the transcription coactivator NPR1 plays dual roles in regulating plant immunity. Cell 137:860
32. Mou Z et al (2003) Inducers of plant systemic acquired resistance regulate NPR1 function through redox changes. Cell 113:935
33. Jung HW et al (2009) Priming in systemic plant immunity. Science 324:89
34. Chanda B et al (2011) Glycerol-3-phosphate is a critical mobile inducer of systemic immunity in plants. Nat Genet 43:421
35. Maldonado AM et al (2002) A putative lipid transfer protein involved in systemic resistance signalling in Arabidopsis. Nature 419:399
36. Heil M, Ton J (2008) Long-distance signaling in plant defense. Trends Plant Sci 13:264
37. Tamogami S et al (2011) Jasmonates to jamolites in plants: past, present, future. Adv Bot Res 60:309
38. Shah J (2009) Plants under attack: systemic signals in defense. Curr Opin Plant Biol 12:459
39. van Loon LC et al (1998) Systemic resistance induced by rhizosphere bacteria. Annu Rev Phytopathol. 36:453
40. Shoresh M et al (2010) Induced systemic resistance and plant responses to fungal biocontrol agents. Annu Rev Phytopathol 48:21

41. Hammerschmidt R (2007) Introduction: definitions and some history. In: Walters DR, Newton A, Lyon G (eds) Induced resistance for Plant defence. Blackwell, Oxford, pp 1–8
42. Vallad GE, Goodman RM (2004) Systemic acquired resistance and induced systemic resistance in conventional agriculture. Crop Sci 44:1920
43. Conrath U (2009) Priming of induced plant defense responses. In: L. C. van Loon (ed) Advances in Botanical Research, vol. 51. Academic Press, pp 361–395
44. Conrath U et al (2006) Priming: getting ready for battle. Mol Plant Microbe Interact 19:1062
45. Kauss H et al (1992) Dichloroisonicotinic and salicylic acid, inducers of systemic acquired resistance, enhance fungal elicitor responses in parsley cells. Plant J 2:655
46. Katz VA et al (1998) A benzothiadiazole primes parsley cells for augmented elicitation of defense responses. Plant Physiol 117:1333
47. Heil M et al (2000) Reduced growth and seed set following chemical induction of pathogen defence: does systemic acquired resistance (SAR) incur allocation costs? J Ecology 88:645
48. Beckers GJM et al (2009) Mitogen-activated protein kinases 3 and 6 are required for full priming of stress responses in *Arabidopsis thaliana*. Plant Cell 21:944
49. Pfluger J, Wagner D (2007) Histone modifications and dynamic regulation of genome accessibility in plants. Curr Opin Plant Biol 10:645
50. Zhang X (2008) The epigenetic landscape of plants. Science 320:489
51. Bannister AJ et al (2002) Histone methylation: dynamic or static? Cell 109:801
52. Conrath U (2011) Molecular aspects of defence priming. Trends Plant Sci 16:524
53. Vaillant I, Paszkowski J (2007) Role of histone and DNA methylation in gene regulation. Curr Opin Plant Biol 10:528
54. Eulgem T, Somssich IE (2007) Networks of WRKY transcription factors in defense signaling. Curr Opin Plant Biol 10:366
55. Pandey SP, Somssich IE (2009) The role of WRKY transcription factors in plant immunity. Plant Physiol 150:1648
56. Rushton PJ et al (2010) WRKY transcription factors. Trends Plant Sci 15:247
57. Jaskiewicz M et al (2011) Chromatin modification acts as a memory for systemic acquired resistance in the plant stress response. EMBO Rep 12:50
58. Luna E et al (2012) Next-generation systemic acquired resistance. Plant Physiol 158:844
59. Pavet V et al (2006) Arabidopsis displays centromeric DNA hypomethylation and cytological alterations of heterochromatin upon attack by *Pseudomonas syringae*. Mol Plant Microbe Interact 19:577
60. Wada Y et al (2004) Association between up-regulation of stress-responsive genes and hypomethylation of genomic DNA in tobacco plants. Mol Genet Genomics 271:658
61. Slaughter A et al (2012) Descendants of primed Arabidopsis plants exhibit resistance to biotic stress. Plant Physiol 158:835
62. Leadbeater A, Staub T (2007) Exploitation of induced resistance: a commercial perspective. In: Walters DR, Newton A, Lyon G (eds) Induced resistance for plant defence. Blackwell, Oxford pp 229–242
63. Midoh N, Iwata M (1996) Cloning and characterization of a probenazole-inducible gene for an intracellular pathogenesis-related protein in rice. Plant Cell Physiol 37:9
64. Kessmann H et al (1994) Induction of systemic acquired disease resistance in plants by chemicals. Annu Rev Phytopathol 32:43
65. Gorlach J et al (1996) Benzothiadiazole, a novel class of inducers of systemic acquired resistance, activates gene expression and disease resistance in wheat. Plant Cell 8:629
66. Tsubata K et al (2006) Development of a novel plant activator for rice diseases, tiadinil. J Pest Sci 31:161
67. Walters DR, Fountaine JM (2009) Practical application of induced resistance to plant diseases: an appraisal of effectiveness under field conditions. J Agric Sci 147:523
68. Chen L et al (2008) A fragment of the *Xanthomonas oryzae* pv. *oryzicola* Harpin HpaGXooc reduces disease and increases yield of rice in extensive grower plantings. Phytopathology 98:792
69. Sharathchandra RG et al (2004) A Chitosan formulation Elexa induces downy mildew disease resistance and growth promotion in pearl millet. Crop Prot 23:881

70. Thanassoulopoulos CC et al (2007) Development of an empirical model to predict losses in eggplant (*Solanum melongena* L.) production caused by Verticillium wilt. Crop Prot 26:08
71. Anderson JP et al (2005) Plant defence responses: conservation between models and crops. Funct Plant Biol 32:21
72. Chern M-S et al (2001) Evidence for a disease-resistance pathway in rice similar to the NPR1-mediated signaling pathway in Arabidopsis. Plant J 27:101
73. Fitzgerald HA et al (2005) Alteration of TGA factor activity in rice results in enhanced tolerance to *Xanthomonas oryzae* pv. *oryzae*. Plant J 43:335
74. Proietti S et al (2011) Cross activity of orthologous WRKY transcription factors in wheat and Arabidopsis. J Exp Bot 62:1975
75. Silverman P et al (1995) Salicylic acid in rice (biosynthesis, conjugation, and possible role). Plant Physiol 108:633
76. Kogel KH et al (1994) Acquired resistance in barley (The resistance mechanism induced by 2,6-dichloroisonicotinic acid is a phenocopy of a genetically based mechanism governing race-specific powdery mildew resistance). Plant Physiol 106:1269
77. Molina A et al (1999) Wheat genes encoding two types of PR-1 proteins are pathogen inducible, but do not respond to activators of systemic acquired resistance. Mol Plant Microbe Interact 12:53
78. Kogel K-H, Langen G (2005) Induced disease resistance and gene expression in cereals. Cell Microbiol 7:1555
79. Shimono M et al (2007) Rice WRKY45 plays a crucial role in benzothiadiazole-inducible blast resistance. Plant Cell 19:2064
80. Takatsuji H et al (2010) Salicylic acid signaling pathway in rice and the potential applications of its regulators. Jpn Agric Res Q 44:217
81. Yu GY, Muehlbauer GJ (2001) Benzothiadiazole-induced gene expression in wheat spikes does not provide resistance to Fusarium head blight. Physiol Mol Plant Pathol 59:129
82. Makandar R et al (2012) Salicylic acid regulates basal resistance to Fusarium head blight in wheat. Mol Plant Microbe Interact 25:431
83. Mayama S et al (1995) Association between avenalumin accumulation, infection hypha length and infection type in oat crosses segregating for resistance to *Puccinia coronata* f. sp. *avenae* race 226. Physiol Mol Plant Pathol 46:255
84. Mayama S, Matsuura Y, Inda H, Tani T (1982) The role of avenalumin in the resistance of oat to crown rust, *Puccinia coronata* f. sp *avenae*. Physiol Plant Pathol 20:189
85. Meydani M (2009) Potential health benefits of avenanthramides of oats. Nutr Rev 67:731
86. Emmons CL, Peterson DM (2001) Antioxidant activity and phenolic content of oat as affected by cultivar and location. Crop Sci 41:1676
87. Wise ML, Doehlert DC, McMullen MS (2008) Association of avenanthramide concentration in oat (*Avena sativa* L.) grain with crown rust incidence and genetic resistance. Cereal Chem 85:639
88. Wise ML (2011) Effect of chemical systemic acquired resistance elicitors on avenanthramide biosynthesis in oat (*Avena sativa*). J Agric Food Chem 59:7028
89. Ren Y, Wise ML (2012) Avenanthramide biosynthesis in oat cultivars treated with systemic acquired resistance elicitors. Cereal Research Comm. doi:10.1556/CRC.2012.0035
90. Wu X et al (2007) Productivity and biochemical properties of green tea in response to full-length and functional fragments of HpaGXooc, a harpin protein from the bacterial rice leaf streak pathogen *Xanthomonas oryzae* pv. *oryzicola*. J Biosci 32:1119
91. Heil M (2007) Trade-offs associated with induced resistance. In: Walters DR, Newton A, Lyon G (eds) Induced resistance for plant defence: a sustainable approach to crop protection. Blackwell, Oxford, pp 157–177
92. Walters DR et al (2011) Possible trade-off associated with the use of a combination of resistance elicitors. Physiol Molec Plant Pathol 75:188
93. Thaler J et al (1999) Trade-offs in plant defense against pathogens and herbivores: a field demonstration of chemical elicitors of induced resistance. J Chem Ecol 25:1597

94. van Hulten M et al (2006) Costs and benefits of priming for defense in Arabidopsis. Nat Acad Sci Proc 103:5602
95. Boyle C, Walters DR (2006) Saccharin-induced protection against powdery mildew in barley: effects on growth and phenylpropanoid metabolism. Plant Pathol 55:82
96. Walters DR et al (2008) Priming for plant defense in barley provides benefits only under high disease pressure. Physiol Mol Plant Pathol 73:95
97. Lucas JA (2011) Advances in plant disease and pest management. J Agric Sci 149(Supplement S1):91
98. Ishii H (2007) Fungicide research in Japan-an overview. 15th International Reinhardsbrunn Symposium on Modern Fungicides and Antifungal Compounds, pp 11–17
99. Walters DR (2010) Induced resistance: destined to remain on the sidelines of crop protection? Phytoparasitica 38:1
100. Stadnik MJ, Buchenauer H (1999) Control of wheat diseases by a benzothiadiazole derivative and modern fungicides. J Plant Dis Protect 106:466
101. Roden LC, Ingle RA (2009) Lights, rhythms, infection: the role of light and the circadian clock in determining the outcome of plant-pathogen interactions. Plant Cell 21:2546
102. Walters DR, Havis ND, Paterson L, Taylor J, Walsh DJ (2011) Cultivar effects on the expression of induced resistance in spring barley. Plant Dis 95:595
103. Reglinski T et al (2007) Integration of induced resistance in crop production. In: Walters DR, Newton A, Lyon G (eds) Induced resistance for plant defence. Blackwell, Oxford, pp 201–228
104. Yi H-S et al (2009) Airborne induction and priming of plant defenses against a bacterial pathogen. Plant Physiol 15:2152
105. Alborn HT et al (1997) An elicitor of plant volatiles from beet armyworm oral secretion. Science 276:945
106. Turlings TCJ et al (2000) Volocitin, an elicitor of maize volatiles in oral secretion of *Spodoptera exigua*: isolation and bioactivity. J Chem Ecol 26:189
107. Engelberth J et al (2004) Airborne signals prime plants against insect herbivore attack. Nat Acad Sci Proc 101:1781
108. Ton J et al (2006) Priming by airborne signals boosts direct and indirect resistance in maize. Plant J 49:16
109. Piesik D et al (2011) Cereal crop volatile organic compound induction after mechanical injury, beetle herbivory (*Oulema* spp.), or fungal infection (*Fusarium* spp.). J Plant Physiol 168:878
110. Degenhardt J (2009) Indirect defense responses to herbivory in grasses. Plant Physiol 149:96

Chapter 5
Phytochemicals for Pest Management: Current Advances and Future Opportunities

Stephen O. Duke, Scott R. Baerson, Charles L. Cantrell, David E. Wedge, Kumudini M. Meepagala, Zhiqiang Pan, Agnes M. Rimando, Kevin K. Schrader, Nurhayat Tabanca, Daniel K. Owens and Franck E. Dayan

Abstract As with pharmaceuticals, a significant proportion of commercial pesticides are natural molecules or are derived from natural compounds. This review describes some of the past commercial successes of phytochemicals as pesticides by pesticide class as well as current work and future prospects for development of pesticides from plant-derived natural compounds. For example, two compounds isolated by assay-guided fractionation of the essential oil of American beautyberry (*Callicarpa americana* L.) (Verbenaceae), callicarpenal and intermediol, were found to have very potent insect repellent properties. An analysis of the number of new phytochemicals being discovered yearly and the relatively few bioassays for potential pesticidal activity that most of the known phytochemicals have been subjected to, indicates that this area still has a bright future. Furthermore, chemical modification of these compounds and their use to discover new modes of action greatly expand the scope for future work. In addition, the use of transgene technology holds great promise, not only to protect crops from pests, by imparting production or manipulation of production of pest management phytochemicals, but also for crop/weed allelopathy, as success in this effort would greatly decrease the most used form of synthetic pesticides, herbicides.

5.1 Introduction

A large fraction of phytochemical secondary compounds owe their existence to the coevolution of the producing plant with its biotic threats, such as herbivorous arthropods and mollusks, plant pathogens, and competing plant species. At least one of their functions in nature is to repel, inhibit, kill, or otherwise avoid damage from

---

S. O. Duke (✉) · S. R. Baerson · C. L. Cantrell · D. E. Wedge · K. M. Meepagala · Z. Pan · A. M. Rimando · K. K. Schrader · N. Tabanca · D. K. Owens · F. E. Dayan
Natural Products Utilization Research, United States Department of Agriculture, Agricultural Research Service, University, MS 38677, USA
e-mail: Stephen.Duke@ars.usda.gov

D. R. Gang (ed.), *50 Years of Phytochemistry Research*,
Recent Advances in Phytochemistry 43, DOI 10.1007/978-3-319-00581-2_5,

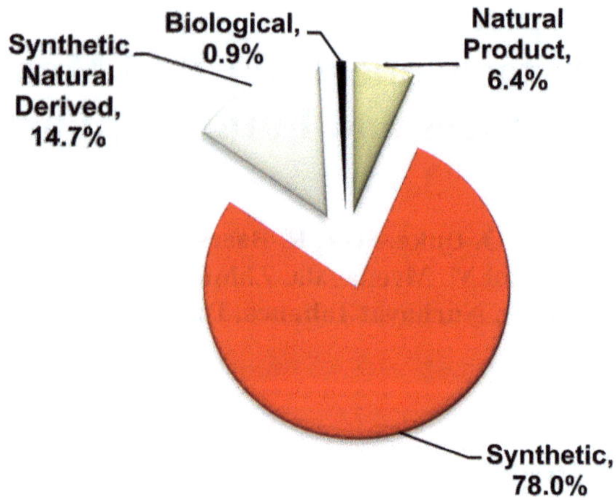

Fig. 5.1 Proportions of pesticides approved by the United States Environmental Protection Agency (USEPA) for use in the USA that are synthetic, natural compounds, derived from natural compounds, and biological agents [1]

these biotic hazards. Thus, these compounds are much more likely to have utility as a pesticide or as a molecular scaffold for pesticide design than compounds in synthetic libraries that have not been designed around compounds with known biological activity. Indeed, as with pharmaceuticals, a significant proportion of commercial pesticides are natural molecules or are derived from natural compounds [1]. From 1997–2010, about 20 % of the new pesticide active ingredients approved for use in the USA were natural products or natural product derivatives (Fig. 5.1). Some of the purely synthetic pesticides were discovered after the discovery of the molecular target site of natural inhibitors. These are not counted in the proportions in Fig. 5.1.

Historically, most of the natural products that have been useful as pesticides or as leads in pesticide discovery have come from plants. Yet, the pesticide industry seems to have focused recently on microbes as sources of leads. Our group spends most of its efforts on discovering potential pesticides from plants. We are also interested in the genetics and synthesis of these compounds as transgene technology allows us to impart production of these natural pesticides into crops.

Several aspects of natural products have reduced interest in them for pesticide discovery. The structural complexity of many natural products is too great for economically feasible production on a commercial scale. Much effort can be wasted in rediscovering known compounds [e.g., 2]. Obtaining enough of some phytochemicals for adequate evaluation can be time consuming and expensive. Sustainable harvest of botanical sources for a compound is often problematic. Natural does not equal nontoxic. We do not cover the mammalian toxicity of the compounds discussed in this short review. Except for materials used in traditional Chinese medicine (TCM), there is very little of this type of information available for the compounds that we mention. The half-lives of many natural compounds are often very short in the environment. This is an environmental advantage, but pesticides must persist sufficiently long to have their desired effects. Patenting can be more complex with natural products for several reasons. Legal complexities with countries or even populations of origin have grown, especially with plant species. This is one reason that even the interest of pharmaceutical companies in phytochemical sources has waned [3]. Lastly, the physicochemical properties of natural compounds are often unsuitable for agricultural use.

Still, there are numerous advantages to phytochemicals in pesticide discovery. They are generally more environmentally benign. They are often sources of new molecular target sites, an aspect that is increasingly important as evolution of pesticide resistance to current modes of action increases [4]. They are often a source of novel chemical structures that differ from those more likely to be devised by traditional pesticide chemists. In some cases, pesticidal phytochemicals have evolved useful selectivity. In addition, modern technology has made discovery of these compounds and their biological activity simpler, faster, and less expensive than a few years ago. Finally, production of these compounds can be transferred from one plant species to another via transgene technology. Crops have sometimes been bred for phytochemical-based pest resistance in the past, but this process has been limited by the phytochemical makeup of related species with which the crop can interbreed.

In this chapter, we briefly describe some of the past commercial successes of phytochemicals as pesticides by pesticide class. Then, we discuss some of the promising work from our group within these pesticide classes.

5.2 Insecticides and Arthropod Repellents

Of all commercial pesticide classes, insecticides have the highest fraction of natural product or natural product-derived products [1]. Slightly more than 30 % of the conventional arthropod pesticides and repellent new active ingredients registered and approved for use by the United States Environmental Protection Agency (USEPA) from 1997 to 2010 were natural products or natural product-derived products [1]. The biggest classes of these compounds, the pyrethroids and the neonicotinoids, originated from phytochemicals. There are also quinoline- and pyrrole-derived commercial insecticides. A very important new class of insecticides are those that target the ryanodine target site in insects. Ryanodine is a compound from the plant *Ryania speciosa* that binds Ca channels of insect muscles [5, 6]. This phytochemical provided the clue for a much-needed new insecticide target site with which to fight evolution of insecticide resistance. Veratridine sulfate from the sabadilla lilly (*Schoenocaulon officinale*) is sold as an insecticide [7]. The scientific literature is full of reports of insecticidal phytochemicals that have not been widely or successfully commercialized.

Several formulations of plant extracts such as neem (*Azadirachta indica*) containing the insect-active compound azadirachtin are sold. There are numerous reviews of neem and azadirachtin as an insecticide [e.g., 8]. Many plant essential oils are available as bioinsecticides, a category of pesticides that does not require the stringent toxicological and environmental testing required of conventional pesticides [9].

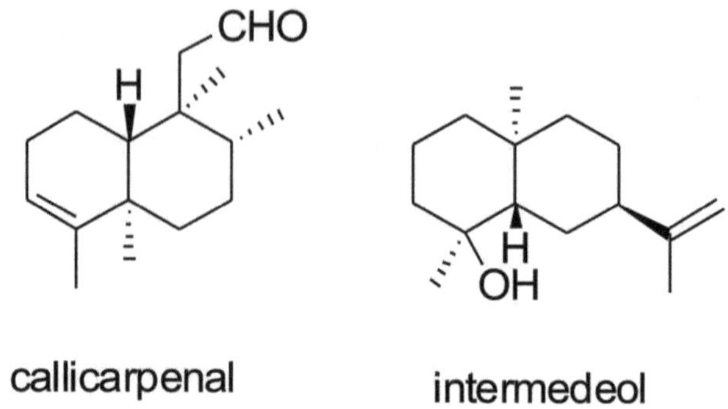

Fig. 5.2 Structures of two insect repellent compounds from *Callicarpa americana*

Our laboratory has focused on insect repellents. One of the more potent repellents is a constituent of the essential oil obtained from American beautyberry (*Callicarpa americana* L.) (Verbenaceae), a common shrub in the US southeast. In Mississippi, crushed leaves of *C. americana* were placed under the harnesses of draft animals as a traditional means to protect the animals from hematophagous insects [10, 11]. Specific identification of the compounds responsible for the mosquito (*Aedes aegypti*) biting deterrence in the leaves of this folk remedy was recently completed using a bioassay-directed fractionation approach. Ultimately, the study identified the compounds callicarpenal and intermediol as those responsible for the biting deterrence from the leaves and hence the folk remedy (Fig. 5.2). Both compounds were evaluated in laboratory bioassays for repellent activity against host-seeking nymphs of the blacklegged tick, *Ixodes scapularis*. Callicarpenal and intermediol, at 155 nmol/cm^2 of cloth, repelled 98 and 96 % of *I. scapularis* nymphs, respectively. Dose–response tests with *I. scapularis* nymphs showed no difference in repellence among callicarpenal, intermediol, and *N, N*-diethyl-*m*-toluamide (DEET) [12]. Callicarpenal, at 155 nmol/cm^2 of cloth, repelled 100 and 53.3 % of *I. scapularis* nymphs at 3 and 4 h, respectively. Both compounds also repel imported fire ants (*Solenopsis* spp.) [13].

More recently, two additional arthropod repellent folk remedies, breadfruit (*Artocarpus altilis*) and *Jatropha* sp., were investigated by Cantrell and colleagues [14, 15]. These two folk remedies are administered traditionally as spatial arthropod repellents by both burning seed-pressed oil in the case of *Jatropha* sp. and burning the dried male inflorescence of breadfruit.

A systematic bioassay-directed study of *Jatropha* sp. oil using adult *Aedes aegypti* females indicated that oleic, palmitic, linoleic, and stearic acids were all active at 25 nmol/cm^2 above a solvent control and were partially responsible for the activity of the oil itself. Evaluation of the triglycerides containing each of these fatty acids revealed that tripalmitin, tristearin, trilinolein, and triolein all demonstrated significant activity above a solvent control at 10 μg/cm^2, with tripalmitin the most active. This study was the first report on the insect repellent activity of triglycerides.

A similar approach to that used for *Jatropha* sp. identified capric, undecanoic, and lauric acids as primary deterrent constituents from the male inflorescence of

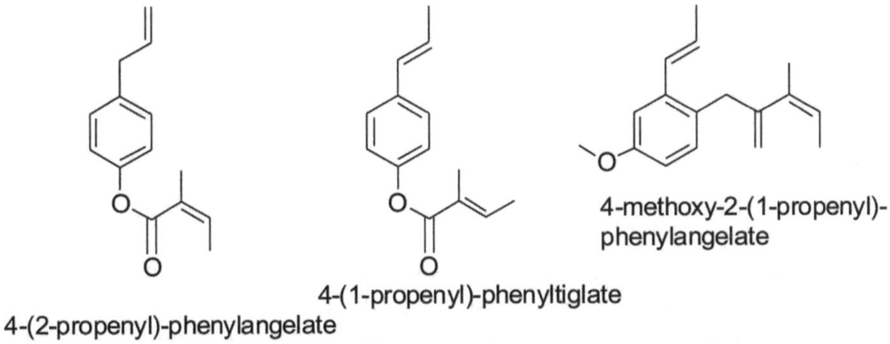

Fig. 5.3 Structures of three major phenylpropanoid constituents of *P. isaurica* oil that were bioassayed against *Liphaphis pseudobrassicae*

breadfruit. A synthetic mixture of fatty acids present in the most active fraction and individual fatty acids was significantly more active than DEET.

Essential oils and plant extracts from plants used in traditional medicines worldwide still continue to provide us with new and unique biological activity. Our group evaluated essential oils from 23 plant species comprising 14 genera and 4 plant families obtained from 26 locations in Turkey [16]. Essential oils obtained by Clevenger distillation were mixed with dimethyl sulfoxide and evaluated for insecticidal activity against adult turnip aphids (*Lipaphis pseudobrassicae* Davis). Aphids were quickly incapacitated by aliphatic aldehydes, phenols, and monocyclic terpenes contained in *Biflora* and *Satureja* species at concentrations as low as 0.3–1.0 mg/ml. *Pimpinella isaurica* essential oil and its three pure phenylpropanoids were tested at a single concentration of 10 mg/ml. Individually, the three major phenylpropanoids—4-(2-propenyl)-phenylangelate, 4-(1-propenyl)-phenyltiglate, and 4-methoxy-2-(1-propenyl)-phenylangelate (Fig. 5.3)—were not toxic to turnip aphids; however, when they were combined they killed aphids. The intact *P. isaurica* oil killed aphids faster than a mixture of the three phenylpropanoids.

We are studying TCM plants to find new agrochemicals with exceptionally low mammalian and environmental toxicity. *Angelica sinensis* (Apiaceae) is one such plant. Dong quai is the Chinese name for the roots of *A. sinensis*, which is a TCM treatment for gynecological disorders. Bioassay-guided fractionation of *A. sinensis* root extract led to the isolation of (*Z*)-ligustilide as an effective insect repellent [17]. This compound had previously been found as an insecticidal constituent of the essential oil of *Ligusticum mutellina* [18]. A mosquito biting deterrence assay showed that (*Z*)-ligustilide (Fig. 5.4) was more potent than the commercial standard DEET to *Ae. Aegypti* and *Anopheles stephensi.*

Essential oils of *Cupressus funebris*, *Juniperus communis*, and *J. chinensis* were evaluated for repellence against adult yellow fever mosquitoes, *Ae. Aegypti*; host-seeking lone star tick nymphs, *Amblyomma amerincanum*; the blacklegged tick, *I. scapularis*, and for toxicity against *Ae. aegypti* larvae and adults [19]. All oils were repellent to both species of ticks. The EC_{95} values of *C. funebris*, *J. communis*,

Fig. 5.4 Structure of a mosquito repellent compound from *Angelica sinensis*

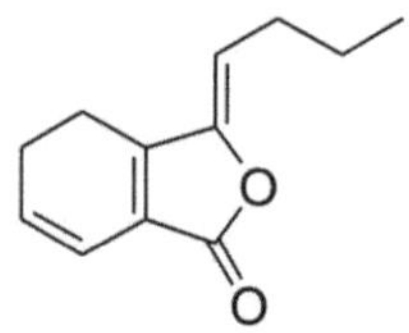

(Z)-ligustilide

and *J. chinensis* oils against *A. americanum* were 0.43, 0.51, and 0.92 mg oil/cm^2 filter paper, respectively, compared to 0.68 mg DEET/cm^2 filter paper. All *I. scapularis* nymphs were repelled by 0.10 mg oil/cm^2 filter paper of *C. funebris* oil. At 4 h after application, 0.83 mg oil/cm^2 filter paper, *C. funebris* and *J. chinensis* oils repelled $\geq$80% of *A. americanum* nymphs. The oils of *C. funebris* and *J. chinensis* did not prevent female *Ae. aegypti* from biting at the highest dosage tested (1.50 mg/cm^2). However, the oil of *J. communis* had a minimum effective dosage (estimate of ED_{99}) for repellence of 0.029 ± 0.018 mg/cm^2; this oil was nearly as potent as DEET. The oil of *J. chinensis* showed a slight ability to kill *Ae. aegypti* larvae, at 80 and 100% at 125 and 250 ppm, respectively.

5.3 Fungicides

Almost 30% of the new fungicide active ingredient registrations in the USA from 1997 to 2010 were either natural products (11.4%) or natural product-derived synthetic compounds (17.1%) [1]. Several of the latter are derived from phytochemicals, such as benzothiazdiazole, acibenzolar-S-methyl, and the alkaloid sanguinarine (Fig. 5.5).

During the past 15 years, we have learned that biological activity of plant extracts against filamentous plant pathogenic fungi does not parallel activity against human pathogenic fungi. We have found that fungicidal chemistry from plants is more common in plants obtained from tropical, moist environments. Medicinal and aromatic plants used in traditional medicine often provide rich sources of novel activity against fungi. A discovery strategy based on this information has led to patenting of sampangine and novel cyclopentenedione compounds for the control of agriculturally important fungal plant pathogens [20, 21].

As part of our ongoing studies on the essential oils, we evaluated *Pimpinella* essential oils that are characterized by high concentrations of pseudoisoeugeneol-type phenylpropanoids. Trinorsesquiterpenes (geijerenes and azulenes) were also found to be characteristic constituents of *Pimpinella* oils [22]. Of the 22 isolated compounds during this investigation, two phenylpropanoids, 4-(3-methyloxiranyl) phenyl 2-methylbutyrate and epoxypseudoisoeugenyl 2-methylbutyrate, showed better antifungal activity than the trinorsesquiterpenes, 4-(6-methylbicyclo[4.1.0]

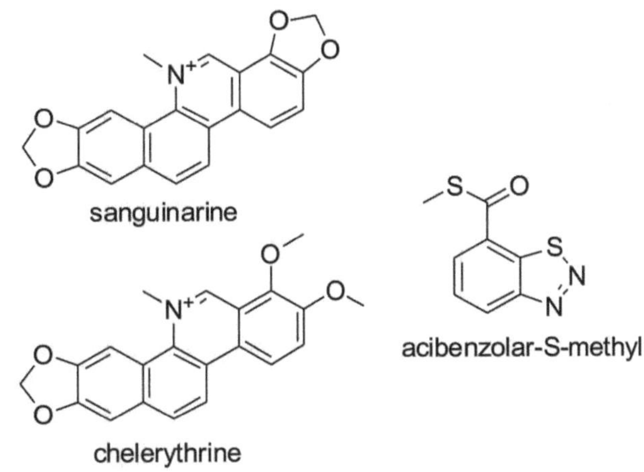

Fig. 5.5 Phytochemical-derived commercial fungicides

hept-2-en-7yl)butan-2-one (tragione) and dictamnol (Fig. 5.6), using direct bioautography against *Collectotrichum acutatum*, *C. fragariae*, and *C. gloesporioides*. The compounds were subsequently evaluated in a 96-well microtiter assay that showed that 4-(3-methyloxiranyl)phenyl 2-methylbutyrate and epoxypseudoisoeugenyl 2-methylbutyrate (Fig. 5.6) produced the most significant growth inhibition in *Phomopsis* spp., *Colletotrichum* spp., and *Botrytis cinerea* [23].

The peanut plant (*Arachis hypogaea* L.), when infected by a microbial pathogen, is capable of producing stilbene-derived compounds that are considered antifungal phytoalexins. In addition, health benefits of some stilbenes from peanuts, including resveratrol and pterostilbene, have been and are being established. Since peanut stilbenoids appear to play roles in plant defense mechanisms, they were evaluated for their effects on economically important plant pathogenic fungi of the genera *Colletotrichum*, *Botrytis*, *Fusarium*, and *Phomopsis*. The results of these studies reveal that peanut stilbenoids, as well as related natural and synthetic stilbene derivatives, display a diverse range of biological activities against fungal plant pathogens [24].

A preparative overpressure layer chromatography (OPLC) method was used for the separation of two new natural compounds, 4-hydroxy-5,6-dimethoxynaphthalene-2-carbaldehyde and 12,13-didehydro-20,29-dihydrobetulin, together with nine known compounds from the acetone extract of the roots of *Diospyros virginiana*. All isolated compounds were evaluated for their antifungal activities against *Colletotrichum fragariae*, *C. gloeosporioides*, *C. acutatum*, *Botrytis cinerea*, *Fusarium oxysporum*, *Phomopsis obscurans*, and *P. viticola* using an *in vitro* micro-dilution broth assay. The results indicated that the compounds methyl-juglone and isodiospyrin (Fig. 5.7) were highly active against *P. obscurans* at 30 μM with 97.0 and 81.4 % growth inhibition, respectively, and moderate activity against *P. viticola* (54.3 and 36.6 %, respectively). OPLC is a rapid and efficient method of exploiting bioactive natural products [25].

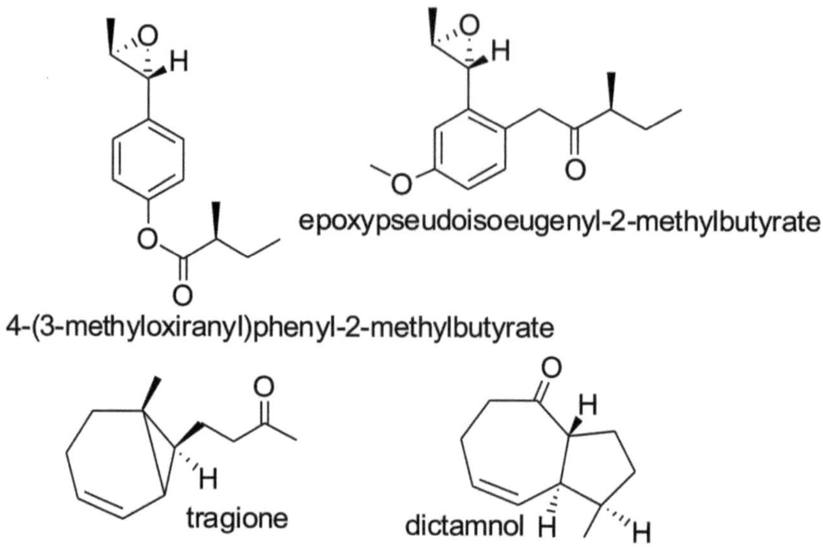

Fig. 5.6 Active antifungal compounds from the essential oil of *Pimpinella* species

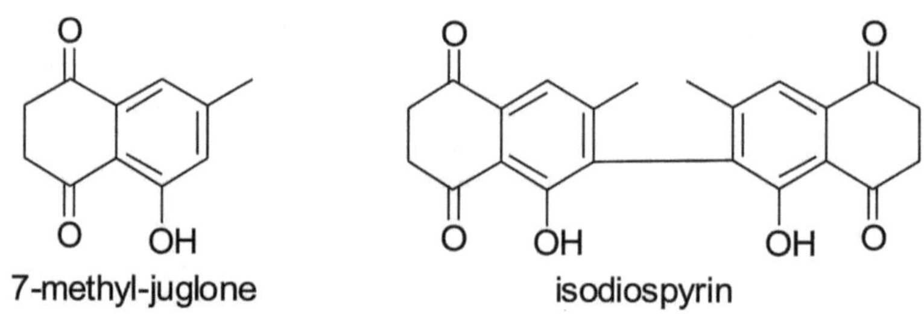

Fig. 5.7 Active antifungal compounds from the roots of *D. virginiana*

5.4 Molluscicides

There are relatively few effective, commercial molluscicides available. However, there are many reports of molluscicidal effects of crude extracts of plants and phytochemicals. For example, a crude butanol extract of *Phytolacca dodecandra* (endod) is effective against *Biomphalaria* snails [26]. Many of these are reviewed by Marston and Hostettmann [27].

Development of the berries of endod as a molluscicide to control schistosomiasis has been very successful in Ethiopia [28]. The active constituents have been isolated and identified as saponins [27]. *Lonicera nigra, Hedera helix, Cornus florida,* and *Asparagus curillus* are among some other plants that have been investigated for molluscicidal saponins [27]. *H. helix* contains a hederagenine glycoside with an LC_{100} of 3 ppm against *B. glabrata* snails.

Our efforts are geared toward development of natural product-based molluscicides to control snails that are harmful for agricultural commodities such as channel catfish (*Ictalurus punctatus*), rice, taro, and orchids. Catfish is one of the main farm-raised fish in the USA. The ram's horn snail (*Planobdella trivolvis*) is an intermediate host for the trematode *Bolbophorus confusus* that was discovered to be a significant problem in 1999 in commercial channel catfish production ponds in the Mississippi Delta region [29, 30], and it has been reported from other states (Arkansas, Louisiana, Alabama, and California). These trematodes have a digenetic life cycle that involves two intermediate hosts, the snail and the catfish, and the American white pelican (*Pelecanus erythrorhynchos*). Catfish infested by the parasitic metacercariae develop cysts, have impaired growth, and are prone to other diseases that can weaken and kill the catfish. The annual economic loss to the catfish industry in the USA due to the trematode problem is estimated to be in millions of dollars. At present, there is no cure or treatment for infected fish. One practical approach to eradicate or control this problem is to interrupt the life cycle of the parasite by eliminating the snails, which are essential to the life cycle.

We have shown that vulgarone B (Fig. 5.8), isolated from the steam distillate of the aerial parts of the plant *Artemisia douglasiana* (Asteraceae), is active toward the snails with an LC_{50} of *ca* 24 μM [31]. The snails showed severe hemolysis associated with lethality when treated with vulgarone B. Channel catfish toxicity studies indicate an LC_{50} of *ca* 207 μM. Thus, vulgarone B may be an environmentally acceptable alternative for snail control in aquaculture when applied within the margin of safety [31]. 2*Z*,8*Z*-matricaria methyl ester (Fig. 5.8) isolated from *Erigeron speciosus* (Asteraceae) has also shown molluscicidal activity against *P. trivolvis* with a LC_{50} of 50 μM associated with marked hemolysis of the snail [32]. In laboratory experiments, Yucca extract at 10 ppm caused 100% mortality of *P. trivolvis*, but ethanol extracts of *Phytolacca americana* (American poke weed) berries and *Lonicera nigra* (black-berried honeysuckle) were inactive (unpublished data). On the other hand, hederagenin 3-O-β-D-glucopyranoside (Fig. 5.8), isolated from English ivy (*H. helix*, an invasive plant in the Northwestern states, showed activity against *P. trivolvis* with an LC_{50} of 30 μM in laboratory studies (unpublished data). This compound has also shown activity against *B. glabrata* snails [27].

The golden apple snail (GAS), *Pomacea canaliculata* (Lamarck), is a major pest of rice in all rice-growing countries outside the USA, where it was either intentionally or accidentally introduced [33]. In the Philippines, the government promoted GAS production for human consumption [34, 35]. However, the demand dropped because GAS was found to transfer the rat lungworm parasite (*Angiostrongylus cantonensis*) to humans if undercooked GAS was consumed. Thus, snail farmers growing GAS abandoned their cultures, and the snails were disposed of without precautions. GAS soon invaded the rice fields, where it found an ideal habitat and abundant food supply. The economic losses were estimated to be up to $ 1.2 billion by 2003 [36]. GAS is also a problem in taro plantations in Hawaii, where $CuSO_4$ is currently used to control the snail population [37].

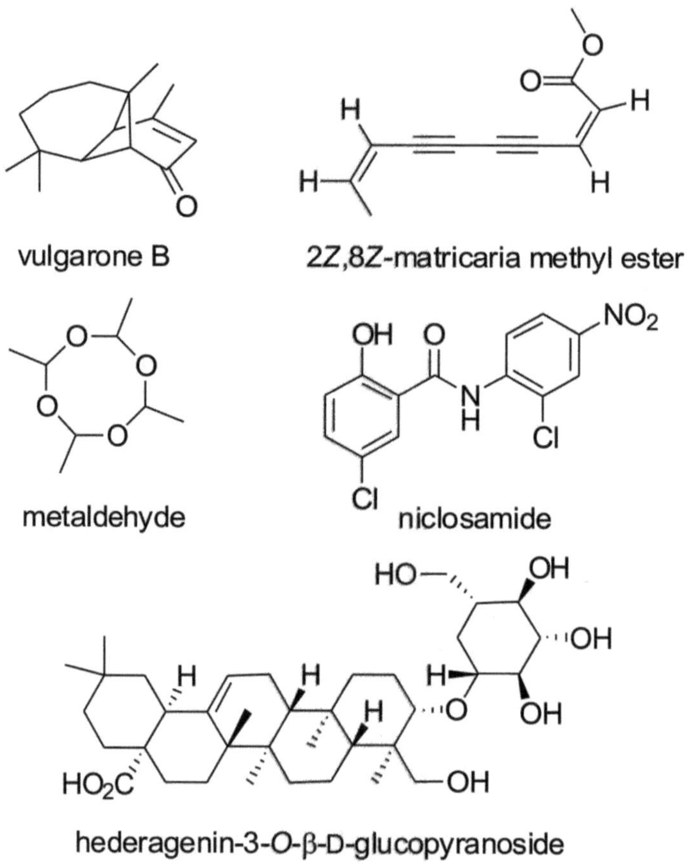

Fig. 5.8 Molluscicidal compounds mentioned in the text

Integrated management methods are recommended for GAS, but many farmers depend on commercially available synthetic molluscicides (niclosamide and metaldehyde—Fig. 5.8). There are numerous cases of poisoning caused by metaldehyde [38, 39]. Therefore, cost-effective, target-specific, and environmentally friendly molluscicides are needed, due to the economic burden and undesirable effects of currently available commercial molluscicides. Vulgarone B is a potential molluscicide with an LC_{50} value of about 30 μM at 24 h for GAS [40]. In the same bioassay, the standard commercial molluscicide, metaldehyde, also had an LC_{50} of about 30 μM. This corresponds to about 6.5 and 4.4 mg/L of the vulgarone B and metaldehyde, respectively. The concentrations needed for 100 % mortality at 24 h were about 75 and 200 μM, respectively, for vulgarone B and metaldehyde. In practical terms, a rice farmer who uses about 250 liters of water for spraying one hectare will require 4.8 g of pure vulgarone B for GAS control [40]. Vulgarone B did not cause further mortality at 48–96 h after treatment, unlike the observed increased mortality

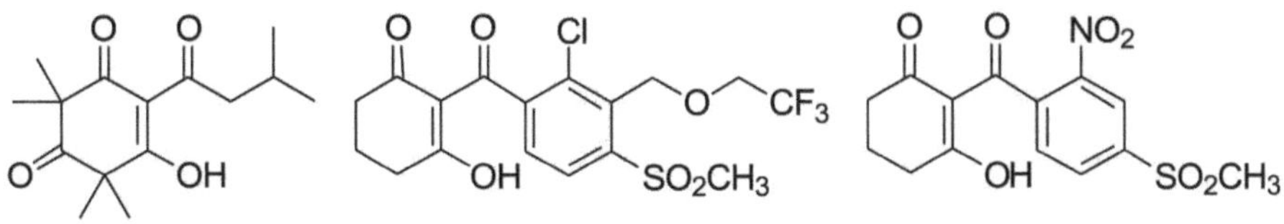

Fig. 5.9 Structures of the natural triketone leptospermone and two synthetic analog that are sold as commercial herbicides

with time with metaldehyde. This indicates that the vulgarone B is fast acting, unlike metaldehyde.

There was no phytotoxicity 10 days after treatment to 14-day-old rice plants at concentrations of vulgarone B that cause complete or nearly complete mortality of GAS. However, when incorporated into agar growth medium, pronounced chlorosis occurred after 14 days of growth. Therefore, vulgarone B should be used after the germination of rice seeds. Field and laboratory experiments have also shown the potential of vulgarone B as a molluscicide in taro paddies in Hawaii (unpublished data).

5.5 Herbicides and Algicides

5.5.1 Herbicides

Only about 8 % of the new active ingredient registrations for conventional herbicides from 1997 to 2010 in the USA have been synthetic, natural product-derived compounds [1]. These have all been triketones that were partially inspired by the plant allelochemical leptospermone (Fig. 5.9) [41]. The discovery and development of these herbicides resulted from a convergence between astute chemical ecology observations made by Reed Gray of the Western Research Center in California (Stauffer Chemical at the time) and independent chemical synthesis efforts in the same laboratory. In 1977, Gray observed that the bottlebrush plant (*Callistemon citrinus*) appeared to repress the growth of plants in its surroundings. Crude extracts from this plant caused the bleaching of grass weeds. He identified the active component as leptospermone, a natural triketone structure with no known biological activity, although it had been reported in a number of Australasian shrubs several years earlier [42]. Leptospermone was moderately active in greenhouse tests, controlling mostly small-seeded grass weeds. This natural product and a small number of synthetic structural analogs were patented as herbicides in 1980 [43]. A few years later, a separate group at the Western Research Center was generating analogs of the cyclohexanedione herbicide sethoxydim, an inhibitor or acetyl-coenzyme-A carboxylase. Some of the second-generation herbicidal derivatives with a dimedone backbone caused bleaching symptoms similar to that from leptospermone. Combination of the syncarpic

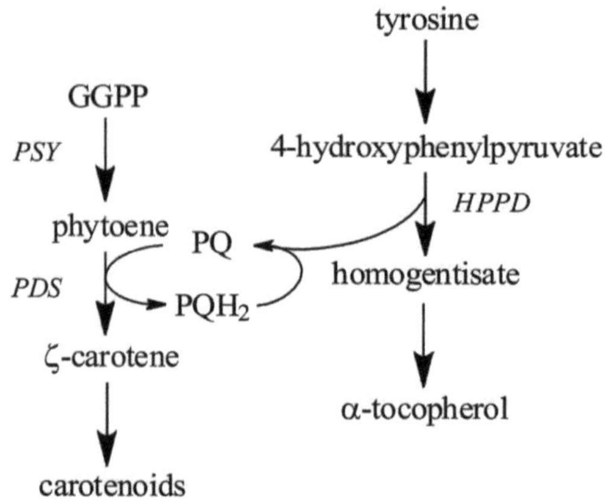

Fig. 5.10 Role of HPPD in carotenoid synthesis. *PSY* phytoene synthase, *PDS* phytoene desaturase *HPPD* *p*-hydroxyphenyl-pyruvate dioxygenase, *PQ* plastoquinone

acid of leptospermone to this chemistry ultimately served as the basis for the development of the triketone synthetic herbicides (Fig. 5.9) [44].

β-Triketones (e.g., leptospermone, flavesone, agglomerone, tasmanone, papuanone, and grandiflorone) are common in many Australasian woody plant genera (e.g., *Leptospermum*, *Eucalyptus*, *Callistemon*, *Xanthostemon*, *Backhousia*, *Calytrix*, *Baeckea*, *Melaceuca*, and *Corymbia*) [42, 45, 46]. On average, steam-distilled manuka oil accounts for 0.3 % of the dry weight of *L. scoparium* [47]. However, the amount of β-triketone present in these oils varies widely across New Zealand. Some chemotypes contain as little as 0.1 % triketone while others can accumulate up to 33 % [47]. More than 200 individual manuka plants from 87 sites throughout New Zealand were analyzed and the triketone-rich chemotypes were almost exclusively limited to the East Cape region [47]. Cluster analysis of the composition of these samples identified 11 geographical chemotypes distinguished by different levels of monoterpenes and sesquiterpenes, methyl cinnamate, and triketones. The reason for this chemotaxonomic geographical distribution is not well understood.

Little is known about the chemotypes outside of New Zealand, though clearly the *Callistemon* samples studied by Gray in 1977 contained sufficient amounts of leptospermone for isolation and purification. Interestingly, this adds to the serendipity of the discovery of its herbicidal activity since the analysis of a number of *Callistemon* species either did not report the presence of detectable amounts of triketones [42] or only trace amounts [48]. The primary constituent of the essential oil of this genus is the monoterpene 1,8-cineole [49].

Synthetic β-triketone herbicides (e.g., sulcotrione and mesotrione) cause bleaching of newly emerging tissues [50]. This symptom was traditionally associated with inhibitors of phytoene desaturase, but triketone herbicides do not inhibit this enzyme. It was later found that these herbicides inhibit *p*-hydroxyphenylpyruvate dioxygenase (HPPD), a key enzyme involved in the biosynthesis of prenyl quinones and tocopherols [50]. Plastoquinone (a prenylquinone) is an essential cofactor for phytoene desaturase [51] (Fig. 5.10). In the absence of plastoquinone, phytoene desaturase activity is reduced which results in the bleaching of young foliage and accumulation of phytoene customarily associated with phytoene desaturase inhibitors [52]. Chlorophyll levels are also affected because the photosynthetic apparatus is no longer protected from reactive oxygen species generated under high light intensity.

Gray observed that leptospermone caused bleaching of plant tissues [43]. Work with the bioactive components of manuka oil demonstrated that some natural β-triketones also inhibit plant HPPD [53]. Most of the activity of manuka oil was attributed to leptospermone because it was the most abundant triketone in the examined oil. However, grandiflorone, a minor constituent that has a more lipophilic side chain, was a much more potent inhibitor of HPPD. Conversely, the short methyl side chain of flavesone nullified the activity of this triketone. The important role of the lipophilicity of the side chain was confirmed by a structure–activity study using a series of natural and synthetic leptospermone analogs [54].

Greenhouse experiments using agricultural soils showed that manuka oil was active both when applied to the foliage and to the soil surface. While most essential oils have little to no soil activity, preemergence application of manuka oil controlled the growth of large crabgrass at a rate of 3 L/ha. The soil activity of manuka oil is due in part to the relatively slow dissipation of leptospermone, which remained active in soil for at least 2 weeks [55].

Triketones and other phytotoxic natural products are often produced and stored in specialized structures, which may serve in part as a mechanism to prevent autotoxic effects [56, 57]. In the leaves of members of the Myrtaceae family, which encompasses most of the known herbicidal triketone-producing species, specialized secretory glands consisting of roughly spherical secretion-filled spaces lined with specialized glandular cells are found (Fig. 5.11). In the genus *Leptospermum*, the gland is typically covered by two to four cells, which have thin, straight walls and are generally of the same approximate size. These cells are encircled by 5–14 unspecialized epidermal cells in a spiral orientation. Although there has been a debate on the method of glandular cavity production, evidence suggests that in the Myrtaceae this formation occurs schizogenously. Schizogenous formation proceeds by the division of single cells within the epidermis or mesophyll layer with the oil cavity forming as an intracellular space [58]. The schizogenous cavity is lined with a single layer of four to six epithelial cells that are thought to be responsible for the production of the volatile oils stored within the cavity [59, 60]. In *Melaleuca* species, the cells lining the immature gland cavity were shown to be metabolically active by osmophilic staining, supporting their role in oil synthesis. In this species, the mature glands contain highly vacuolated epithelial cells lining the gland cavity that are unlikely to lead to continued oil synthesis and accumulation [60]. It has also been demonstrated that essential oil has the potential for release through the modified epidermal cells covering the gland, although the physiological and ecological aspects of this phenomena remain to be investigated.

The *in vitro* chemical synthesis of leptospermone and many other triketones has been well studied [61], but much work remains to unravel the *in vivo* biosynthesis of these compounds. Although an *in planta* biosynthetic route has yet to be established, a hypothetical pathway can be proposed based on the structure of the final compounds (Fig. 5.11). In a series of conversions analogous to the well-examined

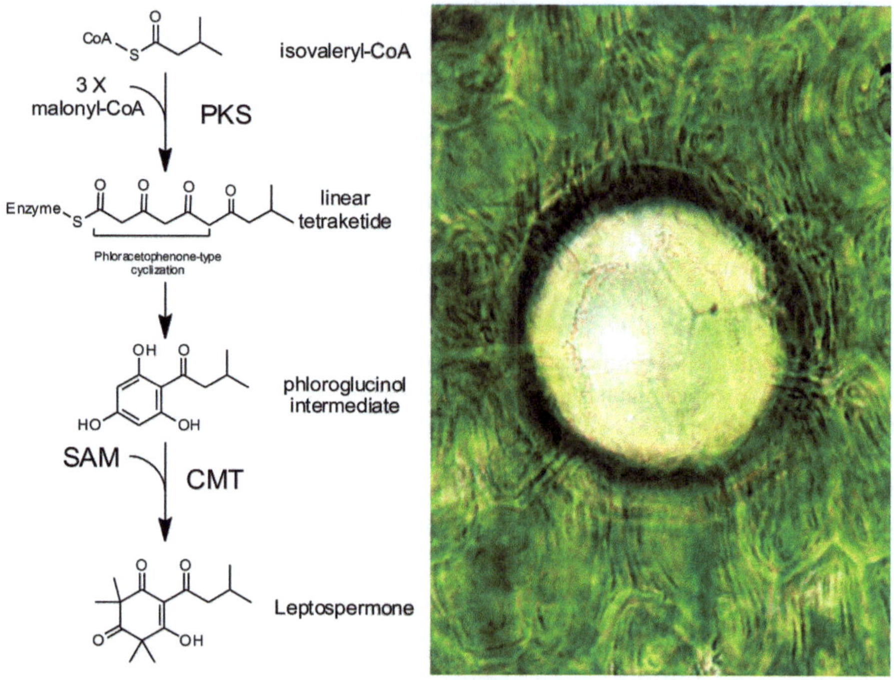

Fig. 5.11 Micrograph of a representative *Leptospermum scoparium* (manuka) schizogenous gland, and a proposed biosynthetic pathway to leptospermone

chalcone synthase enzyme [62], a type III polyketide synthase (PKS) sequentially condenses three malonyl CoA molecules into a polyketide chain extending from an isovaleryl CoA starter molecule. The enzyme subsequently cyclizes the linear tetraketide intermediate via a Claisen-type condensation to generate a phloroglucinol intermediate. A PKS enzyme, valeropenone synthase (VPS), with this activity has been purified to homogeneity and biochemically characterized from *Humulus lupulus* L. (hops) cone glandular hairs [63]. VPS is thought to be involved in the production of the beer flavoring iso-acids of hops which have been shown to contain a β,β-triketone moiety [64]. Subsequently, a gene for this enzyme has been identified and characterized [65]. Efforts are currently underway to isolate and characterize enzymes homologous to VPS from *Leptospermum scoparium* as an initial effort to characterize the leptospermone biosynthetic pathway.

After the production of the phloroglucinol intermediate, the compound would be proposed to undergo spontaneous keto–enol tautomerization, and subsequently to undergo methylation by an as-of-yet unidentified C-methyltransferase (CMT). Early work with methionine-methyl-C^{14} labeled adult *Dryopteris marginalis* ferns demonstrated that the C- and O-methyl substituents of isolated phloroglucinols were derived from methionine [66]. If these findings are consistent with leptospermone, the biosynthetic methyltransferases are likely to be similar to S-adenosylmethionine using CMTs identified in other species.

Table 5.1 Bioassay evaluation results of 9,10-anthraquinone and the analog anthraquinone-59 against the cyanobacterium *Planktothrix perornata* and the green alga *Selenastrum capricornutum*.

Compound	LOEC[a]	LCIC[b]	IC50[c]
P. perornata			
9,10-Anthraquinone	100	100	nd
Anthraquinone-59	10	100	6.3
S. capricornutum			
9,10-Anthraquinone	>100,000	>100,000	nd
Anthraquinone-59	10,000	100,000	5,623

nd not determined
[a] Lowest-observed-effect concentration (nM)
[b] Lowest-complete-inhibition concentration (nM)
[c] 50 % inhibition concentration (nM)

5.5.2 *Algicides*

Some blue-green algae (cyanobacteria) synthesize secondary compounds that can impart unsavory flavors to pond-cultured fish. Currently, aquaculturists are using copper sulfate and chelates, as well as diuron, a synthetic herbicide, to kill cyanobacteria. However, these products generally kill all algae, including beneficial eukaryotic species that are better oxygenators of the water and are not associated with off-flavor compound production. Our laboratory has had a research program to discover natural product-based compounds that are selective for killing cyanobacteria. Among thousands of plant crude extracts and pure compounds tested in the laboratory, 9,10-anthraquinone was one of the most promising compounds [67], with about a thousand times greater activity against a noxious cyanobacterium than on a green alga (Table 5.1). However, the physicochemical properties of this compound were not suitable for use in aquaculture ponds, so the molecule was modified to impart water solubility, while retaining its biological activity [68]. The best modification was the analog anthraquinone-59, which has been patented to control cyanobacteria (Fig. 5.12) [69].

5.6 Transgenic Approaches to Phytotochemical-Based Pest Resistance

All phytochemicals are produced by enzymes encoded by plant genes. With transgenic approaches, we can impart production of new pest management compounds into crops or manipulate the production of such compounds that already exist in crops. Our laboratory has been interested in using these methods to alter the production of sorgoleone in *Sorghum* spp. and other crop species such as rice.

Sorgoleone, a major component of the hydrophobic root exudate of sorghum [*Sorghum bicolor* (L.) Moench], represents one of the most extensively studied

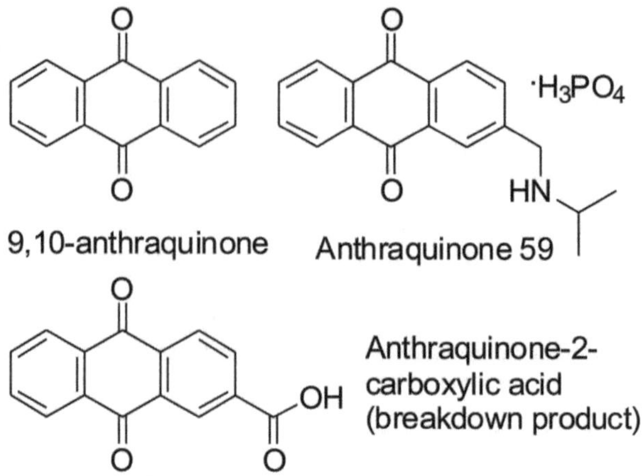

Fig. 5.12 Structures of 9,10-anthraquinone, patented water-soluble analog anthraquinone-59, and the major breakdown product of anthraquinone-59

allelochemicals. In contrast to phenolic lipid-type compounds found in other plants, sorgoleone is an uncommon lipid benzoquinone possessing significant herbicidal activity and is produced exclusively by *Sorghum* spp. [70]. Sorgoleone suppresses the growth of a large number of plant species and appears to be most active on small-seeded plants [71–76]. Additionally, sorgoleone has a relatively long half-life in soil and has been reported to inhibit multiple cellular targets, including plastoquinone/photosystem II, *p*-hydroxyphenylpyruvate dioxygenase, and mitochondrial respiration [77–83]. Thus, evolution of resistance to sorgoleone would presumably be less likely in comparison with a phytotoxin possessing a less-complex mode of action. For the above-mentioned reasons, significant interest has been generated in the potential development of sorgoleone as a natural product-based herbicide.

The sorgoleone biosynthetic pathway appears to exclusively or predominantly reside in root hair cells, with the end product sorgoleone comprising a major portion of the hydrophobic exudate material released into the rhizosphere [80, 84–87]. The biosynthetic pathway of sorgoleone has been previously investigated, and these studies have shown that the aromatic moiety within sorgoleone's structure is synthesized via iterative condensation reactions catalyzed by alkylresorcinol synthase (ARS) enzymes utilizing a C16:3 fatty acyl-CoA precursor [88, 89]. The resulting 5-pentadecatrienyl resorcinol intermediate (produced by ARS) is next methylated by a root hair-specific *O*-methyltransferase [90] and subsequently dihydroxylated by a P450 monooxygenase to yield sorgoleone (Fig. 5.13).

A strategy for the cloning and functional characterization of genes and enzymes involved in sorgoleone biosynthesis has been pursued, involving the screening of candidate gene sequences derived from a root hair-specific *S. bicolor* expressed sequence tag (EST) database [90]. Using this approach, two root hair-specific fatty acid destaurase enzymes were identified, designated SbDES2 and SbDES3, which are likely responsible for the generation of the C16:3 fatty acyl-CoA precursor by consecutive desaturation reactions, starting with palmitoleoyl-CoA (Fig. 5.13) [86]. Heterologous co-expression of *SbDES2* and *SbDES3* in yeast cells resulted in the

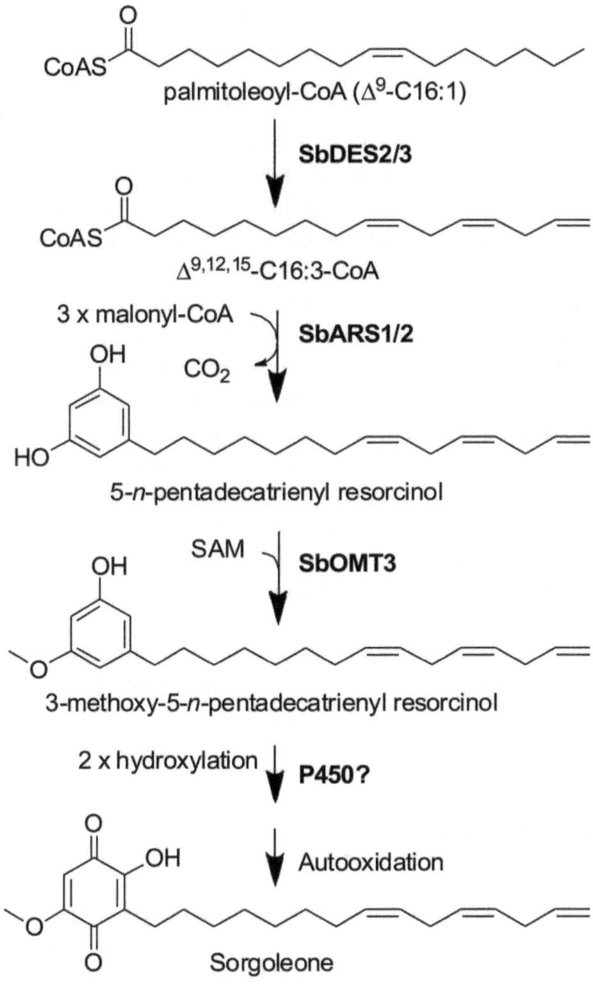

Fig. 5.13 Sorgoleone biosynthetic pathway

production of hexadecatrienoic acid (16:3$\Delta^{9,12,15}$; Fig. 5.13). Co-expression of the two enzymes was required, given that SbDES2 was found to convert endogenous palmitoleic acid (16:1Δ^{9}) to hexadecadienoic acid (16:2$\Delta^{9,12}$), thus providing a substrate that recombinant SbDES3 was capable of converting into hexadecatrienoic acid in vivo [86].

Two root hair-specific ARSs (designated SbARS1 and SbARS2), representing the first such enzymes described from higher plants, have also been characterized and linked to the biosynthesis of sorgoleone [87]. The recombinant SbARS1 and SbARS2 enzymes have both been shown to accept a variety of fatty acyl-CoA starter units using *in vitro* enzymatic assays, including the physiological substrate hexadecatrienyl-CoA (C16:3$\Delta^{9,12,15}$-CoA; Fig. 5.13). The 5-pentadecatrienyl resorcinol intermediate produced by SbARS1 and SbARS2 *in planta* is likely next methylated by a root hair-specific *O*-methyltransferase designated SbOMT3 [90]. Recombinant SbOMT3 was found by *in vitro* enzymatic assays to exhibit a marked preference for alkylresorcinolic substrates among a panel of phenolic compounds tested [90]. As mentioned, the final steps in the sorgoleone biosynthetic pathway involve the dihydroxylation of the 3-methoxy-5-*n*-pentadecatrienyl resorcinol intermediate (Fig. 5.13), and work on several candidate root hair-specific P450 monooxygenase sequences identified within the root hair ESTs is ongoing (Z. Pan, unpublished).

Initial efforts to alter sorgoleone levels in transgenic *S. bicolor* events have been successfully performed with RNA interference (RNAi), using SbARS1/2 3′ coding and contiguous untranslated region (UTR) sequences incorporated within hairpin RNA (hpRNA)-forming binary transformation vectors [87]. For these experiments, segregating T_1 populations representing multiple independent transgenics were first analyzed by quantitative real time polymerase chain reaction (qRT-PCR) for the presence of the transgene-derived hpRNA and individual seedlings were scored as hpRNA "+" or "–". Pooled samples from hpRNA "+" or "–" individuals were analyzed by gas chromatography–mass spectrometry (GC-MS), and sorgoleone levels were found to be reduced to unquantifiable levels in the hpRNA-expressing transformants (Fig. 5.14; see also [87]). The work performed to date on sorgoleone biosynthesis, involving heterologous expression experiments and RNAi in transgenic *S. bicolor*, has provided compelling evidence that SbDES2, SbDES3, SbOMT3, SbARS1, and SbARS2 participate in sorgoleone biosynthesis *in vivo*. These sequences should provide a powerful new toolbox for the manipulation of sorgoleone biosynthesis in *S. bicolor* and the potential transfer of this trait to other crop species.

5.7 Summary

We have provided examples of natural product-based pesticides that are now commercially successful, as well as a few examples of the many natural compounds that we have studied which are active against pests. Our group's research is but a small sampling of the extensive, international effort to discover natural product-based pest management products. Some have argued that we have reached diminishing returns with this approach, but an analysis of the number of new phytochemicals being discovered yearly and the relatively few bioassays for potential pesticidal activity that most of the known phytochemicals have been subjected to, indicates that this is not the case. Furthermore, chemical modification of these compounds and their use to discover new modes of action greatly expands the scope for future work.

It is even clearer that we have only scratched the surface of the possibilities of using transgene technology to protect crops from pests by imparting production or manipulation of production of pest management phytochemicals. We are especially interested in using this method for crop/weed allelopathy [91, 92], as success in this effort would greatly decrease the most used form of synthetic pesticides, herbicides. Furthermore, these efforts provide much-needed experimental verification of the plant/plant allelopathy role of putative allelochemicals. For example, strong support

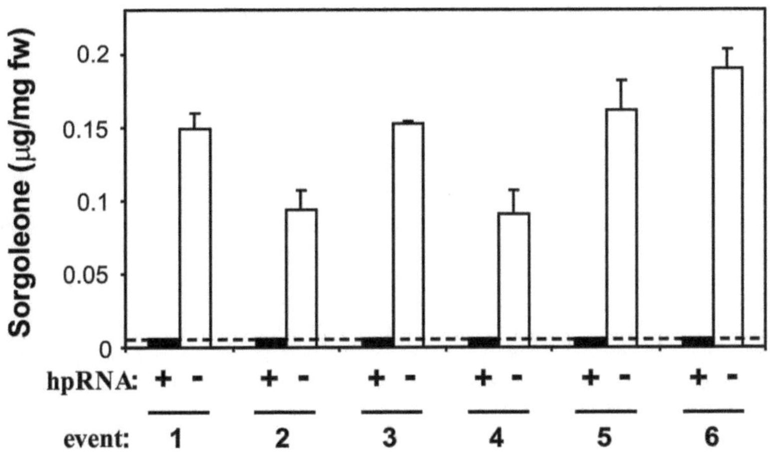

Fig. 5.14 Sorgoleone levels in different *S. bicolor* RNAi transgenic events. Sorgoleone levels were determined by GC-MS analysis of root exudates prepared from pooled hpRNA "+" and hpRNAi "–" seedlings. Data are expressed as mean ± SD

for the allelopathic role of momilactones in rice was recently generated by using gene knockouts to reduce expression of two genes encoding enzymes in the momilactone pathway [93]. These early efforts are promising and should stimulate further research along these lines. Lastly, there can be unexpected spin-off from such research. For example, members of our group have used an *O*-methyltransferase gene of the sorgoleone pathway [90] with a peanut stilbene synthase gene to impart the production of pterostilbene in plants [94]. Pterostilbene is a phytochemical fungicide [e.g., 95], so this technology could be used to improve fungal pathogen resistance in crops. Additional potential benefits of such a transgenic crop are the health-promoting properties of pterostilbene [e.g., 96–98]. Such creative use of the genetics of phytochemical production bodes well for the future of meshing phytochemistry with transgene technology.

References

1. Cantrell CL, Dayan FE, Duke SO (2012) Natural products as sources of new pesticides. J Nat Prod 75:1231–1243
2. Ayer SW, Isaac SK, Krupa DM, Crosby KE, Letendre LJ, Stonard RJ. (1989) Herbicidal compounds from microorganisms. Pestic Sci 27:221–223
3. Ten Kate K, Laird SA (1999) The commercial use of biodiversity. Earthscan Publications, Ltd, London, pp 398
4. Dayan FD, Owens DK, Duke SO (2012) Rationale for a natural products approach to herbicide discovery. Pest Manag Sci 68:519–528
5. Nauen R (2006) Insecticide mode of action: return of the ryanodine receptor. Pest Manag Sci 62:690–692

6. Sattelle DB, Cordova D, Cheek TR (2008) Insect ryanodine receptors: molecular targets for novel pest control chemicals. Invert Neurosci 8:107–119
7. Zhou S, Yin J, Ma J (2011) Control effect of botanical pesticides against *Ectropis oblique hmpulina* and *Empoasa pirisuga* in tea plantation. Plant Dis Pests 2:68–71
8. Mordue AJ, Morgan ED, Nisbet AJ (2005) Azadirachtin, a natural product in insect control. Comp Mol Insect Sci 6:117–135
9. Isman MB (2000) Plant essential oils for pest and disease management. Crop Protection 19:603–608
10. Cantrell CL, Klun JA, Bryson CT, Kobaisy M, Duke SO (2005) Isolation and identification of mosquito bite-deterrent terpenoids from leaves of American (*Callicarpa americana*) and Japanese (*Callicarpa japonica*) beautyberry. J Agric Food Chem 53:5948–5953
11. Krajick K (2006) Keeping the bugs at bay. Science 313:36–38
12. Carrol J, Cantrell CL, Klun J, Kramer M (2007) Repellency of two terpenoid compounds isolated from *Callicarpa americana* (Lamiaceae) against *Ixodes scapularis* and *Amblyomma americanum* ticks. Exp Appl Acarol 41:215–224
13. Chen J, Cantrell CL, Duke SO, Allen ML (2008) Repellency of callicarpenal and intermediol against workers of imported fire ants (Hymenoptera: Formicidae). J Econ Entomol 101:265–271
14. Cantrell CL, Ali A, Duke SO, Khan I (2011) Identification of mosquito biting deterrent constituents from the Indian folk remedy plant *Jatropha curcas*. J Med Entomol 48:836–845
15. Jones AMP, Klun JA, Cantrell CL, Ragone D, Chauhan K, Murch SJ Isolation and identification of mosquito (*Aedes aegypti*) biting deterrent fatty acids from male inflorescences of breadfruit (*Artocarpus altilis* [Partkinson] Fosberg). J Agric Food Chem 60:3867–3873
16. Sampson BJ, Tabanca N, Kirimer N, Demirci B, Baser KHC, Khan IA, Spiers JM, Wedge DE (2005) Insecticidal activity of 23 essential oils and their major compounds against adult *Lipaphis pseudobrassicae* (Davis) (Aphididae: Homoptera). Pest Manag Sci 61:1122–1128
17. Wedge DE, Klun J, Tabanca N, Demirci B, Ozek T, Baser KHC, Liu Z, Zhang S, Cantrell CL, Zhang J (2009) Bioactivity-guided fractionation and GC-MS fingerprinting of *Angelica sinensis* and *A. archangelica* root components for antifungal and mosquito deterrent activity. J Agric Food Chem 57:464–470
18. Passreiter CM, Akhtar Y, Isman MB (2005) Insecticidal activity of the essential oil of *Ligusticum mutellina* roots. Z Naturforsch 60C:411–414
19. Carroll JF, Tabanca N, Kramer M, Elejalde NM, Wedge DE, Bernier UR, Coy M, Becnel JJ, Demirci B, Baser KHC, Zhang J, Zhang S (2011) Activity of *Cupressus funebris*, *Juniperus communis*, and *J. chinensis* (Cupressaceae) essential oils as repellents against ticks (Acari: Ixodidae) and as repellents and toxicants against mosquitoes (Diptera: Culicidae). J Vector Ecol 36:258–268
20. Wedge DE, Nagle DG (2005) Fungicidal properities of sampangine and its analogs to agriculturally important fungal plant pathogens. US Patent 6,844,353 B2, 18 Jan 2005
21. Li XC, Jacob MR, Wedge DE. Novel cyclopentenedione antifungal compounds and methods for their use. US Patent 7,109,380, 19 Sept 2006; USSN: 11/093,695
22. Baser KHC, Tabanca N, Kirimer N, Bedir E, Khan IA, Wedge DE (2007) Recent advances in the chemistry and biological activities of the *Pimpinella* species of Turkey. Pure Appl Chem 79:539–556
23. Tabanca N, Bedir E, Ferreira D, Slade D, Wedge DE, Jacob MR, Khan SI, Kirimer N, Baser HKC, Khan IA (2005) Bioactive constituents from Turkish *Pimpinella* species. Chem Biodivers 2:221–232
24. Sobolev VS, Khan SI, Tabanca N, Wedge DE, Manly SP, Cutler SJ, Coy MR, Becnel JJ, Neff SA, Gloer JB (2011) Biological activity of peanut (*Arachis hypogaea*) phytoalexins and selected natural and synthetic stilbenoids. J Agric Food Chem 59:1673–1682
25. Wang X, Habib E, León F, Radwan MM, Tabanca N, Gao J, Wedge DE, Cutler SJ (2011) Antifungal metabolites from *Diospyros virginiana* by overpressure layer chromatography. Chem Biodivers 8:2331–2340

26. Lemma A, Brody G, Newell GW, Parkhurst RM, Skinner WA (1972) Laboratory evaluation of molluscicidal potency of a butanol extract of *Phytolacca dodecandra* (endod) berries. J Parasitol 58:104–107
27. Marston A, Hostettmann K (1985) Plant molluscicides. Phytochemistry 24:639–652
28. Goll PH, Lemma A, Duncan J, Mazengia B (1983) Control of schistosomiasis in Adwa, Ethiopia, using the plant molluscicide endod (*Phytolacca dodecandra*). Tropenmedizin Parasitologie 34:177–183
29. Pasnik D (1999) Research of new trematode in channel catfish. Fish Farming News. Nov/Dec 40–43
30. Mitchell AJ (2001) Update and impact of a trematode that infects cultured channel catfish. Catfish J 16:17–27
31. Meepagala KM, Sturtz G, Mischke CC, Wise D, Duke S O (2004) Molluscicidal activity of vulgarone B against ram's horn snail (*Planorbella trivolvis*). Pest Manag Sci 60:479–482
32. Meepagala KM, Sturtz G, Wise D, Wedge DE (2002) Molluscicidal and antifungal activity of *Erigeron speciosus* steam distillate. Pest Manag Sci 58:1043–1047
33. Joshi R C, Baucas NS, Joshi EE, Verzola EA (2003) CD-ROM: information database on the golden apple snail (golden kuhol), *Pomacea canaliculata* (Lamarck): a terminal report submitted to the Department of Agriculture-Cordillera Highland Agricultural Resource Management Project. pp 18.
34. Adalla CB, Rejesus BM (1989) The golden apple snail, *Pomacea* sp., a serious pest of lowland rice in the Philippines. In: Henderson I (ed) Slugs and snails in world agriculture. Proceedings of a symposium organized by the British Crop Protection, pp. 417–422
35. Ang W (1984) Snails in human diet. Greenfields 14:30–31
36. Sin TS (2003) Damage potential and control of *Pomacea canaliculata* (Lamarck) in irrigated rice and its control by cultural approaches. Int J Pest Mgmt 49:49–55
37. Hill SA, Miyasaka SC (2000) Taro responses to excess copper in solution culture. HortScience 35:863–867
38. Litsinger JA, Estano DB (1993) Management of the GAS (*Pomacea canaliculata* Lamarck) in rice. Crop Prot 12:363–370
39. Shih CC, Chang SS, Chan Y-L, Chen JC, Chang MW, Tung MS, Deng JF, Yang CC (2004) Acute metaldehyde poisoning in Taiwan. Vet Human Toxicol 46:140–143
40. Joshi RC, Meepagala KM, Sturtz G, Cagauan AG, Mendoza CO, Dayan FE, Duke SO (2005) Molluscicidal activity of vulgarone B from *Artemisia douglasiana* (Besser) against the invasive, alien, mollusc pest, *Pomacea canaliculata* (Lamarck). Int J Pest Manag 51:75–180
41. Knudsen CG, Lee DL, Michaely WJ, Chin H-L, Nguyen NH, Rusay J, Cromartie TH, Gray R, Lake BH, Fraser EM, Cartwright D (2000) Discovery of the triketone class of HPPD-inhibiting herbicides and their relationship to naturally occurring β-triketones. In: Allelopathy in ecological agriculture and forestry. Proceedings of the 3rd International Congress on Allelopathy in Ecological Agriculture and Forestry, pp 101–111
42. Hellyer RO (1968) The occurrence of β-triketones in the steam-volatile oils of some myrtaceous Australian plants. Austral J Chem 21:2825–2828
43. Gray RA, Tseng CK, Rusay RJ (1980) 1-Hydroxy-2-(alkylketo)-4,4,6,6-tetramethyl cyclohexen-3,5-dione herbicides. US Patent 4,202,840
44. Beaudegnies R, Edmunds AJF, Fraser TEM, Hall RG, Hawkes TR, Mitchell G, Schaetzer J, Wendeborn S, Wibley J (2009) Herbicidal 4-hydroxyphenylpyruvate dioxygenase inhibitors—a review of the triketone chemistry story from a Syngenta perspective. Bioorga Medicin Chem 17:4134–4152
45. Douglas MH, van Klink JW, Smallfield BM, Perry NB, Anderson RE, Johnstone P, Weavers RT (2004) Essential oils from New Zealand manuka: triketone and other chemotypes of *Leptospermum scoparium*. Phytochemistry 65:1255–1264
46. van Klink JW, Brophy NB, Perry NB, Weavers RT (1999) Triketones from myrtaceae: Isoleptospermone from *Leptospermum scoparium* and papuanone from *Corymbia dallachiana*. J Nat Prod 62:487–489

47. Douglas M, Anderson R, van Klink J, Perry N, Smallfield B (2001) Defining North Island manuka chemotype resources. Ministry of Agriculture and Forestry Sustainable Farming Fund
48. Brophy JJ, Forster PI, Goldsack RJ, Hibbert DB, Punruckvong A (1997) Variation in *Callistemon viminalis* (Myrtaceae): new evidence from leaf essential oils. Austral Systematic Bot 10:1–13
49. Lassak EV, Smyth MM (1994) Steam volatile leaf oil of *Callistemon linearis* (Schrader et Wendl.) Sweet. J Essential Oil Res 6:403–406
50. Lee DL, Prisbylla MP, Cromartie TH, Dagarin DP, Howard SW, Provan WM, Ellis MK, Fraser T, Mutter LC (1997) The discovery and structural requirements of inhibitors of *p*-hydroxyphenylpyruvate dioxygenase. Weed Sci 45:601–609
51. Norris SR, Barrette TR, DellaPenna D (1995) Genetic dissection of carotenoid synthesis in *Arabidopsis* defines plastoquinone as an essential component of phytoene desaturation. Plant Cell 7:2139–2149
52. Pallett KE, Little JP, Sheekey M, Veerasekaran P (1998) The mode of action of isoxaflutole I. Physiological effects, metabolism, and selectivity. Pestic Biochem Physiol 62:113–124
53. Dayan FE, Duke SO, Sauldubois A, Singh N, McCurdy C, Cantrell CL (2007) *p*-Hydroxyphenylpyruvate dioxygenase is a herbicidal target site for β-triketones from *Leptospermum scoparium*. Phytochemistry 68: 2004–2014
54. Dayan FE, Singh N, McCurdy C, Godfrey CA, Larsen L, Weavers RT, Van Klink JW, Perry NB (2009) β-triketone inhibitors of plant *p*-hydroxyphenylpyruvate dioxygenase: modeling and comparative molecular field analysis of their interactions. J Agric Food Chem 57:5194–5200
55. Dayan FE, Howell JL, Marais JM, Ferreira D, Koivunen ME (2011) Manuka oil, a natural herbicide with preemergence activity. Weed Sci 59:464–469
56. Duke SO, Rimando AM, Duke MV, Paul RN, Ferreria JFS, Smeda RJ. (1999) Sequestration of phytotoxins by plants: Implications for biosynthetic production. In: Cutler HA, Cutler SJ (eds) Natural products: agrochemicals and pharmaceuticals. CRC Press, Boca Raton, pp 127–136
57. Dayan FE, Duke SO (2003) Trichomes and root hairs: natural pesticide factories. Pesticide Outlook 4:175–178
58. Carr DJ, Carr SGM (1970) Oil glands and ducts in *Eucalyptus* l'herit. II. Development and structure of oil glands in the embryo. Austral J Bot 18:191–212
59. Fahn A (1988) Secretory tissues in vascular plants. New Phytol 108:229–257
60. List S, Brown PH, Walsh KB (1995) Functional anatomy of oil glands of *Melaleuca alternifolia* (Myrtaceae). Austral J Bot 43:629–641
61. Singh IP, Sidana J, Bharate SB, Foley WJ (2010) Pholoroglucinol compounds of natural origin: synthetic aspects. Nat Prod Rep 27:393–416
62. Austin MB, Noel JP (2003) The chalcone synthase superfamily of type III polyketide synthases. Nat Prod Rept 20:79–110
63. Paniego NB, Zuurbier KWM, Fung SY, van der Heijden R, Schefeer, JJC, Verpoorte R (1999) Phlorisovalerophenone synthase, a novel polyketide synthase from hop *(Humulus lupulus* L.) cones. Eur J Biochem 262:612–616
64. Hughes P (2000) The significance of iso-α-acids for beer quality. J Instit Brewing 106:271–276
65. Okada Y, Ito K (2001) Cloning and analysis of valerophenone synthase gene expressed specifically in lupulin gland of hop (*Humulus lupulus* L.). Biosci Biotechnol Biochem 65:50–155
66. Penttila A, Kapadia GJ, Fales HM (1965) The biosynthesis *in vivo* of methylenebisphloroglucinol derivatives. J Amer Chem Soc 87:4402–4403
67. Schrader KK, de Regt MQ, Tidwell PR, Tucker CS, Duke SO (1998) Selective growth inhibition of the musty-odor producing cyanobacterium *Oscillatoria* cf. *chalybea* by natural compounds. Bull Environ Contamination Toxicol 60:651–658
68. Schrader KK, Nanayakkara NPD, Tucker CS, Rimando AM, Ganzera M, Schaneberg BT (2003) Novel derivatives of 9,10-anthraquinone are selective algicides against the musty-odor cyanobacterium *Oscillatoria perornata*. Appl Environ Microbiol 69:5319–5327
69. Schrader KK, Nanayakkara NPD (2005) Selective algai-cides for control of cyanochloronta. US Patent 6,949,250, September 2005

70. Dayan FE, Rimando AM, Pan Z, Baerson SR, Gimsing A-L, Duke SO (2010) Molecules of interest: Sorgoleone. Phytochemistry 71:1032–1039
71. Barbosa LCA, Ferreira ML, Demuner AJ, da Silva AA, de Cassia Pereira R (2001) Preparation and phytotoxicity of sorgoleone analogues. Quim Nova 24:751–755
72. de Souza CN, de Souza IF, Pasqual M (1999) Extração e ação de sorgoleone sobre o crescimento de plantas. Ciênc Agrotechnol 23:331–338
73. Einhellig FA, Souza IF (1992) Phytotoxicity of sorgoleone found in grain sorghum root exudates. J Chem Ecol 18:1–11
74. Forney DR, Foy CL, Wolf DD (1985) Weed suppression in no-till alfalfa (*Medicago sativa*) by prior cropping with summer-annual forage grasses. Weed Sci 33:490–497
75. Netzly DH, Butler LG (1986) Roots of sorghum exude hydrophobic droplets containing biologically active components. Crop Sci 26:775–778
76. Panasiuk O, Bills DD, Leather GR (1986) Allelopathic influence of *Sorghum bicolor* on weeds during germination and early development of seedlings. J Chem Ecol 12:1533–1543
77. Rasmussen JA, Hejl AM, Einhellig FA, Thomas JA (1992) Sorgoleone from root exudate inhibits mitochondrial functions. J Chem Ecol 18:197–207
78. Einhellig FA, Rasmussen JA, Hejl AM, Souza IF (1993) Effects of root exudates sorgoleone on photosynthesis. J Chem Ecol 19:369–375
79. Gonzalez VM, Kazimir J, Nimbal C, Weston LA, Cheniae GM (1997) Inhibition of a photosystem II electron transfer reaction by the natural product sorgoleone. J Agric Food Chem 45:1415–1421
80. Czarnota MA, Paul RN, Dayan FE, Nimbal CI, Weston LA (2001) Mode of action, localization of production, chemical nature, and activity of sorgoleone: a potent PSII inhibitor in *Sorghum* spp. root exudates. Weed Technol 15:813–825
81. Weston LA, Czarnota MA (2001) Activity and persistence of sorgoleone, a long-chain hydroquinone produced by *Sorghum bicolor*. J Crop Prod 4:363–377
82. Meazza G, Scheffler BE, Tellez MR, Rimando AM, Nanayakkara NPD, Khan IA, Abourashed EA, Romagni, JG, Duke SO, Dayan FE (2002) The inhibitory activity of natural products on plant *p*-hydroxyphenylpyruvate dioxygenase. Phytochemistry 59:281–288
83. Demuner LAJ, Barbosa CA, Luiz S, Chinelatto LS, Reis C (2005) Sorption and persistence of sorgoleone in red–yellow latosol. Quim Nova 28:451–455
84. Czarnota MA, Rimando AM, Weston LA (2003) Evaluation of root exudates of seven sorghum accessions. J Chem Ecol 29:2073–2083
85. Dayan FE, Watson SB, Nanayakkara NPD (2007) Biosynthesis of lipid resorcinols and benzoquinones in isolated secretory plant root hairs. J Exp Bot 58:3263–3272
86. Pan Z, Rimando AM, Baerson SR, Fishbein M, Duke SO (2007) Functional characterization of desaturases involved in the formation of the terminal double bond of an unusual $16{:}3\Delta^{9,12,15}$ fatty acid isolated from *Sorghum bicolor* root hairs. J Biol Chem 282:4326–4335
87. Cook D, Rimando AM, Clemente TE, Schröder J, Dayan FE, Nanayakkara N, Pan Z, Noonan BP, Fishbein M, Abe I, Duke SO, Baerson SR (2010) Alkylresorcinol synthases expressed in *Sorghum bicolor* root hairs play an essential role in the biosynthesis of the allelopathic benzoquinone sorgoleone. Plant Cell 22:867–887
88. Fate GD, Lynn DG (1996) Xenognosin methylation is critical in defining the chemical potential gradient that regulates the spatial distribution in *Striga* pathogenesis. J Am Chem Soc 118:11369–11376
89. Dayan FE, Kagan IA, Rimando AM (2003) Elucidation of the biosynthetic pathway of the allelochemical sorgoleone using retrobiosynthetic NMR analysis. J Biol Chem 278:28607–28611
90. Baerson SR, Dayan FE, Rimando AM, Nanayakkara NPD, Liu C-J, Schröder J, Fishbein M., Pan Z, Kagan IA, Pratt LH, Cordonnier-Pratt M-M, Duke SO (2008) A functional genomics investigation of allelochemical biosynthesis in *Sorghum bicolor* root hairs. J Biol Chem 283:3231–3247
91. Duke SO, Scheffler BE, Dayan FE, Ota E (2001) Strategies for using transgenes to produce allelopathic crops. Weed Technol 15:826–834

92. Duke SO, Baerson SR, Rimando AM, Pan Z, Dayan FE, Belz RG (2007) Biocontrol of weeds with allelopathy: conventional and transgenic approaches. In Vurro M, Gressel J (eds) Novel biotechnologies for biocontrol agent enhancement and management. Springer, Dordrecht, pp 75–85
93. Xu M, Galhano R, Wiemann P, Bueno E, Tiernan M, Wu W, Chung I-M, Gershenzon J, Tudzynski B, Sesma A, Peters RJ (2012) Genetic evidence for natural product-mediated plant-plant allelopathy in rice (*Oryza sativa*). New Phytol 193:570–575
94. Rimando AM, Pan Z, Polaschock JJ, Dayan FE, Mizuno C, Snook ME, Liu C-J, Baerson SR (2012) *In planta* production of the highly potent resveratrol analogue pterostilbene via stilbene synthase and *O*-methyltransferase co-expression. Plant Biotech J 10:269–283
95. Slaughter AR, Hamiduzzaman MM, Neuhaus J-M, Mauch-Mani M (2008) Beta-aminobutyric acid-induced resistance in grapevine against downy mildew: involvement of pterostilbene. Eur J Plant Pathol 122:185–195
96. Joseph JA, Fisher DR, Cheng V, Rimando AM, Shukitt-Hale B (2008) Cellular and behavioral effects of stilbene resveratrol analogues: implication for reducing the deleterious effects of aging. J Agric Food Chem 56:10544–10551
97. Rimando AM, Nagmani R, Feller DR, Yokoyama W (2005) Pterostilbene, a new agonist for the peroxisome proliferator-activated receptor α-isoform, lowers plasma lipoproteins and cholesterol in hypercholesterolemic hamsters. J Agric Food Chem 53:3403–3407
98. Paul S, Rimando AM, Lee HJ, Ji Y, Reddy B, Suh N (2009) Anti-inflammatory action of pterostilbene is mediated through the p38 mitogen-activated protein kinase pathway in colon cancer cells. Cancer Prevention Res 2:650–657

Chapter 6
Sulfhydryl-Reactive Phytochemicals as Dual Activators of Transcription Factors NRF2 and HSF1

Ying Zhang, Sharadha Dayalan Naidu, Rumen V. Kostov, Ashley Pheely, Vittorio Calabrese and Albena T. Dinkova-Kostova

Abstract Two central regulators, nuclear factor-erythroid 2 p45-related factor 2 (NRF2) and heat shock factor 1 (HSF1), control the KEAP1/NRF2/ARE pathway and the heat shock response, two essential cellular defense mechanisms. Both systems are highly inducible under conditions of stress. Many small molecules, including certain phytochemicals, such as isothiocyanates and phenylpropanoids, and/or their metabolites, have the capacity to induce the KEAP1/NRF2/ARE pathway. Recent results suggest that a common signal that is sensed through cysteine modification(s) within Kelch-like ECH-associated protein 1 (KEAP1) and HSF1, or possibly within a negative regulator of HSF1, is responsible for triggering both pathways. Celastrol, withaferin A, gedunin, curcumin, and sulforaphane are examples of structurally diverse phytochemicals with a common chemical signature: reactivity with sulfhydryl groups. This reactivity underlies their biological activities as multitarget agents for which protective effects have been documented in numerous animal models of human disease and which include induction of large networks of transcriptional programs regulated by transcription factors NRF2 and HSF1.

A. T. Dinkova-Kostova (✉) · Y. Zhang · S. D. Naidu · R. V. Kostov · A. Pheely
Division of Cancer Research, Medical Research Institute, University of Dundee, Dundee DD1 9SY, Scotland, UK
e-mail: A.DinkovaKostova@dundee.ac.uk

V. Calabrese
Department of Chemistry, University of Catania, 95100 Catania, Italy

A. T. Dinkova-Kostova
Departments of Medicine and Pharmacology and Molecular Sciences, Johns Hopkins University School of Medicine, Baltimore, MD, USA

D. R. Gang (ed.), *50 Years of Phytochemistry Research*,
Recent Advances in Phytochemistry 43, DOI 10.1007/978-3-319-00581-2_6,

6.1 Introduction

The KEAP1/NRF2/ARE pathway and the heat shock response represent two essential cellular defense mechanisms that are controlled by two central regulators, nuclear factor-erythroid 2 p45-related factor 2 (NRF2) and heat shock factor 1 (HSF1), respectively. Both systems do not normally operate at their maximal capacity, but are highly inducible under conditions of stress. Induction leads to the transcriptional upregulation of networks of proteins that protect against the potentially devastating consequences of thermal, oxidative, and electrophilic stress, and chronic inflammation. Nearly 25 years ago, Paul Talalay and his colleagues discovered that many small molecules, including certain phytochemicals, such as isothiocyanates and phenylpropanoids, and/or their metabolites, have the capacity to induce the KEAP1/NRF2/ARE pathway. Although structurally very diverse, excluding the possibility of ligand–receptor interactions as the underlying mechanism, all inducers possess a common chemical signature: reactivity with sulfhydryl groups [1]. Based on this finding, it was predicted that there exists in the cell a protein sensor for inducers that is endowed with highly reactive cysteine residues [2], which was later identified by Masayuki Yamamoto and his colleagues as Kelch-like ECH-associated protein 1 (KEAP1) [3], the main negative regulator of transcription factor NRF2 [4]. More recently, reactivity with sulfhydryl groups has emerged as also being important for the activation of HSF1 by various small molecules [5–9]. Thus, it appears that a common signal that is sensed through cysteine modification(s) within KEAP1 and HSF1, or possibly within a negative regulator of HSF1, is responsible for triggering both pathways.

6.2 The KEAP1/NRF2/ARE Pathway

The KEAP1/NRF2/ARE pathway is at the forefront of the cellular defense. In numerous experimental systems, induction of this pathway has been shown to be protective against various conditions of stress. Conversely, failure to upregulate the pathway (such as under conditions of NRF2 deficiency) leads to increased sensitization and accelerated disease pathogenesis. Under basal conditions, transcription factor NRF2 is continuously targeted for ubiquitination and proteasomal degradation by the repressor protein KEAP1, which serves as a substrate adaptor for Cullin 3 (Cul3)-based E3 ubiquitin ligase (Fig. 6.1) [10–12]. In addition to KEAP1, the levels of NRF2 within the cell are also controlled by the action of glycogen synthase kinase-3β (GSK3β) and β-transducin repeat-containing protein (β-TrCP) which serves as a substrate adaptor for Cullin 1 (Cul1)-based ubiquitin ligase [13]. The precise mechanistic details of regulation of the KEAP1/NRF2/ARE pathway are not completely understood and several different models have been proposed [14]. It is clear, however, that many inducers of the pathway chemically modify specific cysteine residues within KEAP1, leading to loss of its ability to target NRF2

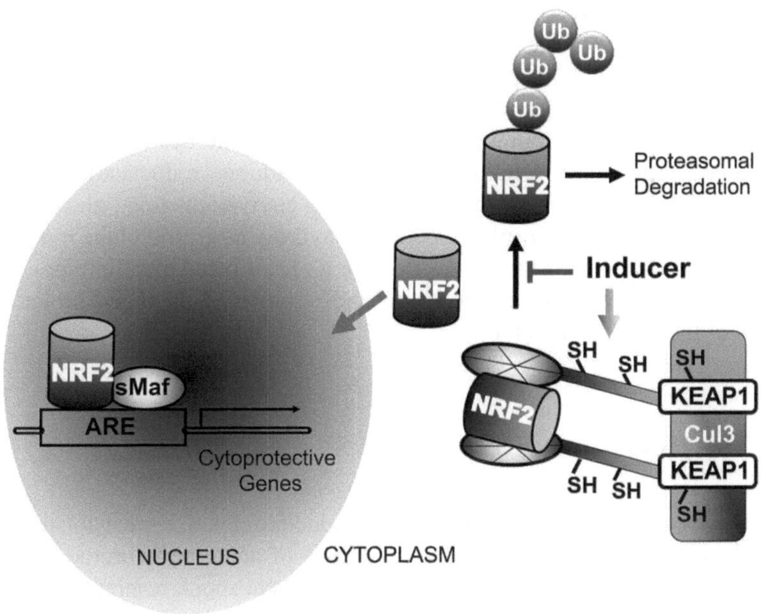

Fig. 6.1 The KEAP1/NRF2/ARE pathway. Under basal conditions, NRF2 is targeted for ubiquitination and proteasomal degradation by its repressor KEAP1, which serves as a substrate adaptor for Cullin 3 (Cul3)-based ubiquitin ligase. Inducers chemically react with cysteine residues of KEAP1, rendering it unable to target NRF2 for degradation. As a result, NRF2 accumulates and undergoes nuclear translocation, where it binds to antioxidant response elements (AREs) as a heterodimer with a small Maf protein, driving the expression of cytoprotective genes

for ubiquitination and proteasomal degradation. Subsequently, NRF2 accumulates, enters the nucleus, binds as a heterodimer with a small Maf transcription factor to antioxidant response elements (AREs, specific sequences that are present in the promoter regions of NRF2-target genes), and activates transcription [15–17].

NRF2-dependent genes encode a large network of cytoprotective proteins, including those that are involved in the metabolism and transport of a wide array of endo- and xenobiotics, proteins that have antioxidant functions, as well as those that participate in the synthesis, utilization, and regeneration of glutathione and NADPH. The number of genes that are under the transcriptional control of NRF2 is fascinatingly large: a recent study integrating chromatin-immunoprecipitation with parallel sequencing (ChIP-Seq) and global transcription profiling identified 645 basal and 654 inducible direct targets of NRF2, with 244 genes at the intersection [18]. Moreover, the functional diversity of the NRF2-dependent cytoprotective proteins is extraordinary and provides the cell with multiple layers of protection. Examples of NRF2-dependent proteins include: (1) antioxidant enzymes (e.g., heme oxygenase 1 (HO-1), NAD(P)H:quinone oxidoreductase 1 (NQO1), and thioredoxin reductase); (2) conjugating enzymes (e.g., glutathione *S*-transferases (GSTs) and uridine 5′-diphospho (UDP)-glucuronosyltransferases); (3) proteins that facilitate the export of xenobiotics and/or their metabolites (e.g., solute carriers and adenosine triphosphate (ATP)-binding cassette transporters); (4) anti-inflammatory enzymes (e.g., leukotriene B_4 dehydrogenase); (5) enzymes that participate in the synthesis and regeneration of glutathione (e.g., χ-CT, the core subunit of the cystine/glutamate membrane transporter, γ-glutamate cysteine ligase catalytic (GCLC) and modulatory (GCLM) subunits, glutathione reductase); (6) enzymes that are responsible for the synthesis of reducing equivalents (e.g., glucose 6-phosphate dehydrogenase, 6-phosphogluconate dehydrogenase, and malic enzyme); (7) proteins that protect against metal overload (e.g., ferritin and metallothioneins); and (8) proteins that participate in the repair and removal of damaged proteins (e.g., subunits of the 26S proteasome).

6.3 The Heat Shock Response

The heat shock response is another critical inducible defense mechanism which is essential in protecting the cell under conditions of acute and chronic proteotoxic stress affecting the integrity of the proteome. The heat shock response is controlled by a family of heat shock (transcription) factors, among which HSF1 plays the major role [19, 20]. Under homeostatic conditions, HSF1 is an inactive monomeric phosphoprotein bound to Hsp90 (Fig. 6.2). Following stimulation, HSF1 dissociates from the Hsp90 complex, trimerizes, and binds to heat shock elements (HSEs) of its target genes, thereby driving their expression [19–23]. In addition, a number of posttranslational modifications, such as phosphorylation, sumoylation, and acetylation, are involved in regulating the transcriptional activity of HSF1, and there is also negative feedback regulation by heat shock proteins, such as Hsp70 and Hsp40. Several different ways of activation of HSF1 have been proposed and the experimental evidence for each one of them was recently reviewed [23]. Displacement of HSF1 from its negative regulator Hsp90 is one major mechanism: indeed, pharmacological inhibition of Hsp90 or its antibody-mediated depletion is sufficient to induce trimerization and DNA binding of HSF1 [24, 25].

Similarly to the KEAP1/NRF2/ARE pathway, the number of genes that are regulated by the heat shock response is strikingly large: various studies employing differential display, transcriptional profiling, or proteomic approaches have shown that, depending on the organism, approximately 50–200 genes are induced [26]. According to their functions, the proteins encoded by these genes have been grouped into seven distinct classes: (1) molecular chaperones that prevent unspecific aggregation of nonnative or partially misfolded proteins (e.g., Hsp70, Hsp40); (2) proteolytic proteins that participate in the removal of irreversibly damaged proteins (e.g., BAG3 (BCL2-associated protein), APG5 L (protein involved in autophagy), the cysteine protease, caspase 1 (CASP1), neural precursor cell-expressed developmentally downregulated 4 like (NEDD4 L), and ubiquitin-protein ligase)); (3) RNA- and DNA-modifying enzymes, which are necessary to repair DNA damage (e.g., the bacterial DNA glycosylase MutM); (4) metabolic enzymes that are needed

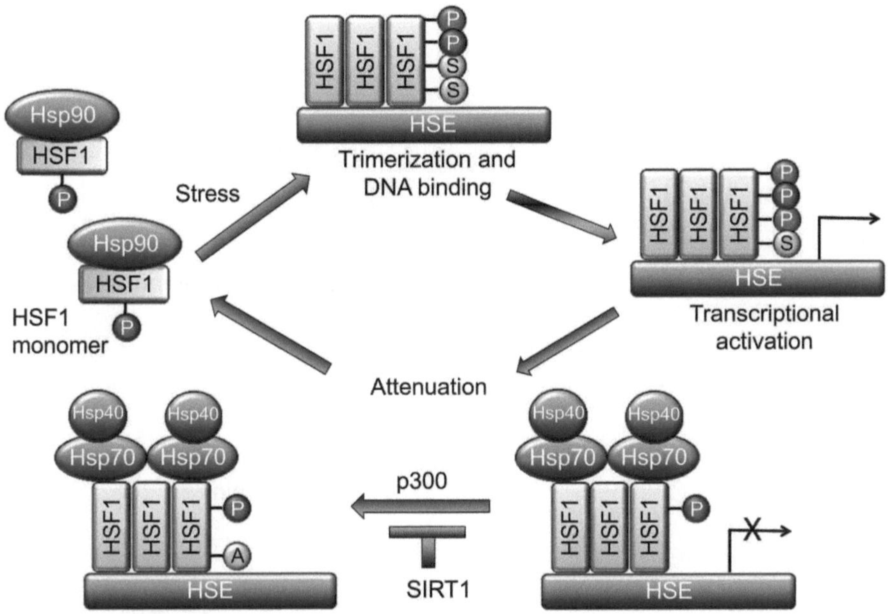

Fig. 6.2 The heat shock response. Under basal conditions, HSF1 is an inactive monomeric phosphoprotein bound to Hsp90. Following stimulation, HSF1 dissociates from Hsp90, trimerizes, and binds to heat shock elements (HSEs) of its target genes, thereby driving their expression. In addition, multiple posttranslational modifications regulate the transcriptional activity of HSF1, such as phosphorylation (P), sumoylation (S), and acetylation (A). HSF1 is also negatively feedback-regulated by heat shock proteins. (Adapted from [19])

to reorganize and maintain the energy supply of the cell (e.g., ACAT2 (acetyl-CoA acetyltransferase), ALAS1 (aminolevulinate synthase), ChGn (chondroitin α-1,4-*N*-acetylgalactosaminyltransferase)); (5) transcription factors, kinases, or phosphatases that further activate other stress response pathways (e.g., RHOH (Ras homolog), PTPG1 (tyrosine phosphatase), RGS2 (regulator of G-protein signaling), and IER5 (regulator of immediate early response)); (6) proteins involved in sustaining cellular structures such as the cytoskeleton and membranes (e.g., tight junction-associated protein (TJP4) and signal-induced proliferation-associated 1-like protein 3 (SIPA1L3)); and (7) proteins involved in transport and detoxification (e.g., the amino acid transporter SLC38A2).

It should be emphasized that the KEAP1/NRF2/ARE pathway and the heat shock response are two distinct defense mechanisms. Thus, induction of the KEAP1/NRF2/ARE pathway occurs in the absence of HSF1; likewise, induction of the heat shock response is independent of NRF2 [9]. Nevertheless, there is some functional overlap between the two pathways which is perhaps best exemplified by HO-1, also known as Hsp32. Indeed, the gene encoding HO-1 is one of the most highly inducible genes (in terms of both kinetics and magnitude of induction) in response to both heat shock as well as inducers of the KEAP1/NRF2/ARE pathway.

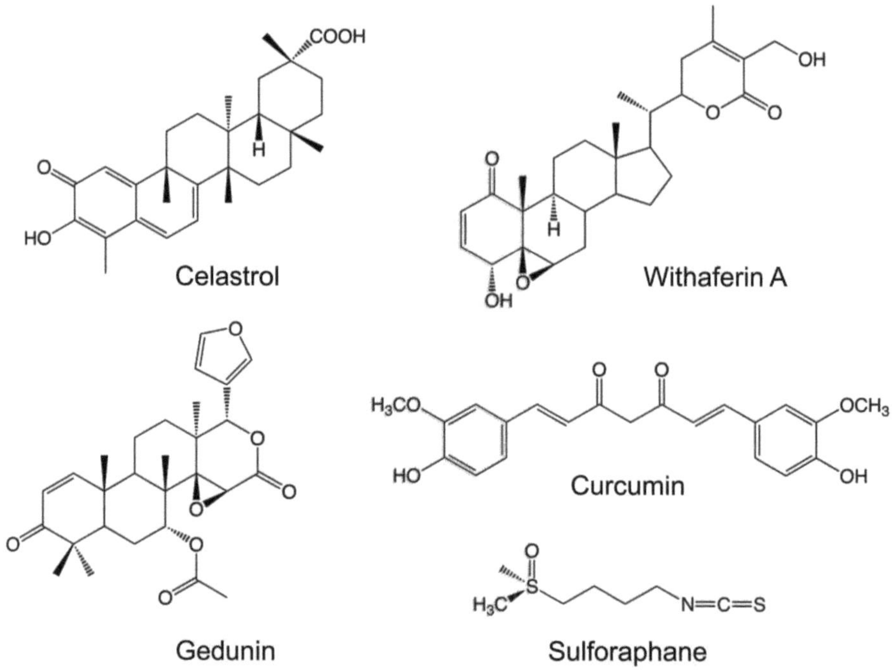

Fig. 6.3 Examples of sulfhydryl-reactive phytochemicals—dual activators of transcription factors NRF2 and HSF1

6.4 Phytochemicals that Activate Transcription Factors, NRF2 and HSF1, and the Consequences of Activation

6.4.1 Celastrol

The quinone methide triterpenoid celastrol (Fig. 6.3) was isolated from the Chinese plant *Tripterygium wilfordii*. Celastrol was reported to induce the heat shock response following a screen of bioactive small molecules in the human HeLa cell line hsp70.1pr-luc [27]. This cell line is stably transfected with a luciferase-encoding construct under the transcriptional control of the *hsp70* promoter. The same study showed that celastrol activates the *hsp70* promoter reporter in several different cell types (i.e., in the breast cancer cell lines MCF7 and BT474, the non-small-cell lung carcinoma cell line H157, and the neuroblastoma cell line SH-SY5Y) to levels comparable with those induced by heat shock (42 °C). Treatment with celastrol led to hyperphosphorylation of HSF1, enhanced binding of HSF1 to the heat shock element in the Hsp70 promoter, and transcriptional activation of endogenous heat shock genes.

Expression profiling of RNA isolated from the androgen-dependent prostate cancer cell line LNCaP that had been treated with celastrol was performed in order to obtain a gene expression signature for celastrol activity [28]. A collection of gene expression profiles of drug-treated cell lines, termed the Connectivity Map [29], was then used to identify known drugs with similar effects on gene expression. Strikingly, the celastrol gene expression signature was found to be very similar to those of four known inhibitors of Hsp90, strongly suggesting that celastrol functions as an inhibitor of Hsp90. This conclusion was further supported by the ability of celastrol to: (1) decrease the levels of the Hsp90 client proteins AR, FLT3, EGFR, BCL-ABL1, AKT, and HER-2 in a concentration-dependent manner in several different cell lines [28, 30], (2) inhibit the ATP-binding activity of Hsp90 [28], and (3) reduce the interaction of Hsp90 with the co-chaperone p23 [28, 31].

Treatment with celastrol was found to disrupt the interaction of Hsp90 and cell division cycle protein 37 (Cdc37), a co-chaperone which is essential for the association of client proteins to Hsp90 [32]. As a result, the Hsp90 client proteins Cdk4 and AKT were destabilized and degraded via the ubiquitin proteasome, and apoptosis was initiated in the pancreatic cancer cell line Panc-1. Further mechanistic studies revealed that celastrol reacts with the C-terminus of Hsp90 and inhibits the ATPase activity of the chaperone without affecting the ATP binding pocket [33]. To identify target proteins of celastrol, stable isotope labeling with amino acids in cell culture (SILAC) approach was used by Hansen et al. [34] in cultured human lymphoblastoid cells that had been exposed to celastrol for 24 hours. It was found that 158 of the ~1,800 proteins with robust quantitation had at least a 1.5-fold change in their levels, with 112 being upregulated and 46 being downregulated. Upregulated proteins include those involved in cellular homeostatic processes, stress responses, cell death, and intracellular transport. A prominent group is that involved in protein quality control, such as the endoplasmic reticulum molecular chaperones GRP78 (HspA5), Grp94 (Hsp90B1), calnexin (CANX), calreticulin (CALR), ERp29 (ERP29), multiple protein disulfide isomerases, glucosidases, and glycosyltransferases. A second group of celastrol-induced proteins comprises those involved in the cellular defense against oxidative stress, such as peroxiredoxins, thioredoxins, and HO-1. These findings are in agreement with an earlier study by Trott et al. [5] in *Saccharomyces cerevisiae* in which transcriptional profiling showed that treatment with celastrol induced heat shock genes as well as antioxidant genes. Celastrol caused hyperphosphorylation of the yeast HSF1 and upregulation of heat shock proteins. In addition, transcription factor Yap1, which is activated in response to oxidants and electrophiles and triggers the transcription of cytoprotective genes, was also activated by celastrol treatment, via the carboxy-terminal redox center of the transcription factor. Similar to its effects in yeast, celastrol also induces antioxidant response genes (e.g., GCLM, χ-CT, and NQO1), in parallel with heat shock target genes (e.g., Hsp70) in RKO human colorectal carcinoma cells [5], Hepa1c1c7 mouse hepatoma cells, and mouse embryonic fibroblasts [9]. Induction of Hsp70 requires functional HSF1, but is independent of NRF2, whereas upregulation of NQO1 occurs in the absence of HSF1, but the presence of NRF2 is essential [9].

The ability of celastrol to upregulate the KEAP1/NRF2/ARE pathway and the heat shock response suggests that celastrol could have cytoprotective effects, a conclusion that has received experimental support in a number of different studies.

One protective effect of celastrol is its ability to inhibit endogenous peroxynitrite formation and to prevent endothelial barrier dysfunction [35]. Celastrol was also shown to protect against aminoglycoside-induced hair cell death and to reduce hearing loss in mice receiving systemic aminoglycoside treatment [36]. Exposure to celastrol was protective against lethal heat stress in both HeLa cells and SH-SY5Y cells, and to a similar extent as a 42° C heat shock [37]. In cells ectopically expressing a mutant polyglutamine (Q57-YFP) protein, celastrol treatment prevented the aggregation of the fusion protein and the associated cytotoxicity [37]. In rodent models for Alzheimer's disease, celastrol reduced the amyloid-α-associated pathology [38] and improved memory, learning, and psychomotor activity [39]. In mice, celastrol protects against 1-methyl-4-phenyl-1,2,3,6-tetrahydropyridine (MPTP)- and 3-nitropropionic acid-induced neurotoxicity [40]. In a transgenic mouse model of amyotrophic lateral sclerosis (ALS), celastrol prevented neuronal cell death, improved weight loss and motor performance, and delayed disease onset [41].

In addition to its cytoprotective effects, celastrol has also been shown to inhibit the proliferation of a number of cancer cell lines and to suppress tumor development and metastasis in various animal models of carcinogenesis. Thus, treatment of the oral leukoplakia cell line MSK-Leuk1 with celastrol inhibited the activation of AhR-dependent transcription of *CYP1A1* and *CYP1B1*, which encode proteins that are responsible for the conversion of polycyclic aromatic hydrocarbons to genotoxic metabolites [42]. Consequently, the formation of benzo(a)pyrene [B(a)P]-induced DNA adducts was reduced. The antitumor activity of celastrol has also been demonstrated using a panel of human breast cancer cell lines with selectivity toward those overexpressing the receptor tyrosine kinase ErbB2, an Hsp90 client protein [43]. Furthermore, celastrol inhibited tumor growth of ErbB2-overexpressing human breast cancer cells in a mouse xenograft model [43]. Celastrol downregulated the expression of CXCR4, a chemokine receptor that is closely linked with tumor metastasis, in breast cancer, colon cancer, squamous cell carcinoma, and pancreatic cancer cells, and inhibited invasion [44]. Reduction of hypoxia-induced angiogenesis and metastasis by celastrol has also been demonstrated and shown to be partly mediated by inhibition of Hsp90 [45]. Hsp90 inhibition was also implicated in the ability of celastrol to increase the sensitivity of the NCI-H460 lung cancer cell line to radiation [46]. In the RIP1-Tag2 transgenic mouse model of pancreatic islet carcinoma, tumor metastasis was suppressed by more than 80 % when celastrol was administered at a dose of 3 mg/kg body weight, once every 3 days, for 4 weeks [32].

6.4.2 Gedunin

Gedunin (Fig. 6.3) is a tetranortriterpenoid isolated from the Indian neem tree *Azadirachta indica* which has antimalarial, insecticidal, and anticancer activity. Using the Connectivity Map [29], it was found that, similar to celastrol-, gedunin-induced genes were enriched in the profile of known Hsp90 inhibitors; likewise, gedunin-repressed genes were repressed by known Hsp90 inhibitors [28]. Although less potent, gedunin appears to share with celastrol some of the same mechanisms by which it inhibits Hsp90 and activates NRF2. Thus, gedunin treatment caused a decrease in ATP binding to Hsp90 and a reduction of the protein levels of the Hsp90 client proteins AR, FLT3, EGFR, and BCR-ABL1 in a concentration-dependent manner in three different cell lines: LNCaP, K562, and Ba/F3 [28]. Similarly, when MCF-7 cells were incubated with gedunin for 24 h, there occurred a dose-dependent degradation of the Hsp90 client proteins HER-2 and AKT, strongly suggesting that Hsp90 is the cellular target for this compound [30]. Indeed, gedunin has been shown to disrupt the interaction between Hsp90 and its co-chaperone, Cdc37 [30, 47].

Recently, gedunin was identified as an NRF2-dependent inducer of cytoprotective enzymes in a high-throughput screen of the Spectrum library comprising 2,000 biologically active compounds [48]. This screening assay used a Neh2-luciferase reporter system in which the Neh2 domain of NRF2, through which the transcription factor binds to its negative regulator KEAP1, was fused to firefly luciferase, thus allowing the direct monitoring of induction based on the time course of reporter activation. The same study showed that gedunin protected neurons against oxidative stress in an astrocyte-dependent manner, and via an NRF2-dependent mechanism. Thus, when primary cultured astrocytes were pretreated with gedunin for 24 h followed by the addition of neurons in the presence of the glutathione-depleting compound homocysteic acid, significant neuroprotection was observed. Protection by gedunin was accompanied by an increase in the levels of GSH and HO-1, and was abrogated by silencing of NRF2.

Gedunin was also shown to induce apoptosis and inhibit cell growth in Caco-2 (colon cancer), MCF-7 (breast adenocarcinoma), NCI-H460 (non-small-cell lung cancer), and A375-C5 (melanoma) cells [49, 50]. In the SKOV3, OVCAR4, and OVCAR8 ovarian cancer cell lines, treatment with gedunin decreased cell proliferation by 80 % [51]. When a bioinformatics analysis was performed, integrating gedunin sensitivity and gene expression data from 54 cancer cell lines, 52 genes were found to be associated with gedunin sensitivity, 49 of which had decreasing expression with increasing gedunin sensitivity [51]. Pathway analysis revealed significant alterations in signaling pathways controlled by the aryl hydrocarbon receptor, phosphatidylinositol 3-kinase (PI3K)/AKT, nitric oxide, neuregulin, and extracellular signal-regulated kinase/mitogen-activated protein kinase. In agreement with the protective effects of gedunin, three closely related compounds, deoxygedunin, deacetoxy-7-oxogedunin, and deacetylgedunin, were shown to activate HSF1 and induce Hsp70 [52]. Moreover, in an MG-132-induced protein misfolding neuronal cell culture model, the compounds protected against cell death, and RNAi knockdown of HSF1 significantly reversed the cytoprotective effect. In HeLa cells that had been transiently transfected with a polyglutamine-expanded toxic isoform (Q103) of huntingtin, a model for Huntington's disease, all three gedunin derivatives improved cell survival.

6.4.3 Withaferin A

Withaferin A (Fig. 6.3) is a withanolide found in the Indian medicinal plant *Withania somnifera*. More than 30 withanolides have been reported to induce the NRF2-dependent enzyme NQO1 [53, 54]. Withaferin A was among the active compounds which were recently identified as inducers of the HSF1-dependent heat shock response when a library of more than 80,000 natural and synthetic compounds were evaluated using a reporter cell line optimized for high-throughput screening [8]. This reporter cell line expresses enhanced green fluorescent protein (EGFP) under the transcriptional control of a minimal consensus HSE-containing promoter. Withaferin A was also shown to bind to the C-terminus of Hsp90, disrupt the Hsp90–Cdc37 complex, inhibit the activity of the chaperone, and promote the degradation of the Hsp90 client proteins AKT, Cdk4, and the glucocorticoid receptor [55]. Molecular docking analyses of withaferin A into the active Hsp90–Cdc37 complex support the hypothesis that this withanolide has the potential to inhibit the association of the chaperone with its co-chaperone by disrupting the stability of attachment of Hsp90 to Cdc37 [56, 57].

The anticancer effects of withaferin A have been demonstrated in various cell culture and xenograft models. In MCF-7 cells, withaferin A downregulated estrogen receptor alpha (ERα), and caused apoptosis and cell growth inhibition [58]. Withaferin A induced apoptosis and inhibited cell proliferation in the pancreatic cancer cell lines Panc-1, MiaPaCa2, and BxPc3 [55] and in the glioma cell lines LN428, LN827, U87, and BT70, irrespective of their p53 status [8]. Furthermore, in xenograft models of pancreatic Panc-1 cells, tumor growth was reduced by withaferin A in a dose-dependent manner [55]. Tumor growth inhibition by withaferin A was also demonstrated in an orthotopic xenograft model in mice which involved intracranial implantation of BT70 glioma progenitor cells [8]. In this model, evaluation of the mRNA levels for HO-1 within the intracranial tumor mass showed a dramatic (7.7-fold) increase in the withanolide A-treated animals.

6.4.4 Curcumin

Curcumin (1,7-bis(4-hydroxy 3-methoxy phenyl)-1,6-heptadiene-3,5-dione, diferuloylmethane; Fig. 6.3) is a component of turmeric, a yellow spice that is obtained from the rhizomes of the East Indian plant *Curcuma longa* L. (Zingiberaceae). We have described the ability of curcumin and other related phenolic Michael acceptors to induce the NRF2-dependent enzyme NQO1 in Hepa1c1c7 cells [59, 60]. Curcumin has also been shown to increase the expression of HO-1 in rat neurons and astrocytes [61], renal epithelial cells [62], and human cardiac myoblasts, hepatocytes, monocytes, and endothelial cells [63–66]. In the human proximal tubule cell line HK-2, curcumin increased the expression of Hsp70 and protected against shiga toxin-induced apoptosis and necrosis [67]. Curcumin treatment increased the levels of HO-1 and Hsp70, and improved the functional recovery of pancreatic islets following cryopreservation [68]. In cultured *Xenopus laevis* A6 kidney epithelial cells, curcumin led to induction of Hsp30 and Hsp70 and was protective against a subsequent cytotoxic thermal challenge [69]. In the chronic myelogenous leukemia cell line K562, exposure to curcumin disrupted the binding the Hsp90–p23 complex to its client protein p210 BCR/ABL, and thus decreased the levels of this oncogenic tyrosine kinase [70]. Hsp27, Hsp70, and Hsp40 were induced when lung adenocarcinoma (CL1–5) cells were treated with curcumin [71].

The hepatic enzyme activities of catalase and superoxide dismutase as well as the levels of Hsp70 were increased when curcumin was administered intravenously to Sprague-Dawley rats [72]. This treatment protected the liver against the damaging effects of ischemia/reperfusion, including lipid peroxidation, activation of inducible nitric oxide synthase and myeloperoxidase, and apoptosis, and improved survival. Pretreatment with curcumin enhanced induction of heat shock proteins Hsp70, Hsp27, and alpha B crystalline in liver and adrenal gland of rats that had been subjected to heat stress [73]. Numerous studies have demonstrated the protective effects of curcumin in animal models of neurodegeneration, cardiovascular disease, diabetes, and cancer; these have been recently reviewed [74–76]. In humans, curcumin is well tolerated at doses up to 12 g per day [77]. As of October 2013, curcumin has been or currently is in 85 different clinical trials targeting various disease conditions, such as psoriasis, radiation dermatitis, atopic asthma, mild cognitive impairment, Alzheimer's disease, ulcerative colitis, multiple myeloma, pancreatic cancer, colorectal cancer, and myelodysplastic syndrome (www.clinicaltrials.gov).

6.4.5 *Sulforaphane*

Sulforaphane (1-isothiocyanato-(4*R*)-(methylsulfinyl)butane; Fig. 6.3) is an isothiocyanate that is formed upon plant tissue injury from a glucosinolate precursor (glucoraphanin), a phytochemical present in cruciferous plants. Sulforaphane was isolated from extracts of broccoli (*Brassica oleracea*) as the principal inducer of the NRF2-dependent enzyme NQO1 using a highly quantitative bioassay in murine hepatoma Hepa1c1c7 cells [78]. Over the past 20 years, voluminous experimental evidence that is beyond the scope of this *Perspective* has convincingly shown the ability of sulforaphane to induce NRF2-dependent genes in cells and animals, and to protect against chronic degenerative conditions, such as cancer, cardiovascular disease, and neurodegenerative diseases (reviewed in [79–81]). Induction of HSF-1 dependent genes by this isothiocyanate is a more recent discovery. Global gene expression profiling of liver tissue isolated from C57BL/6J mice that had been treated with a single dose of sulforaphane (90 mg/kg, *p.o.*) revealed that, in addition to the classical NRF2-dependent genes, there was a prominent induction of heat shock proteins, including alpha B crystallin, Hsp40, and Hsp70 [82]. In HL60 and K562 cells, two human leukemic cell lines that have a chromosome abnormality known as the Philadelphia chromosome, sulforaphane treatment was reported to cause nuclear accumulation of both NRF2 and HSF1, and to increase the expression of Hsp70; interestingly, these effects were enhanced following hGSTA1-1 overexpression [83]. Activation of HSF1 and the heat shock response was also shown in human HeLa and monkey COS-1 cells, with increased expression in Hsp27 being implicated in upregulation of the proteasomal activity [84].

In a pancreatic cancer xenograft model, co-treatment with sulforaphane enhanced the antitumor effect of the 17-allylamino 17-demethoxygeldanamycin (17-AAG), an Hsp90 inhibitor, resulting in inhibition of tumor growth by more than 70% [85]. The isothiocyanate disrupted the Hsp90–Cdc37 interaction and, synergistically with 17-AAG, downregulated several Hsp90 client proteins, including mutant p53, Raf-1, and Cdk4 [85]. Thus, inhibition of Hsp90 may underlie the ability of sulforaphane to activate HSF1 and the heat shock response. In addition to disrupting the association of Hsp90 with its co-chaperone, another potential mechanism of sulforaphane inhibiting the activity of Hsp90 is through altering the acetylation of the chaperone. Sulforaphane was discovered to downregulate the activity of histone deacetylase (HDAC) in a number of human cell lines established from colon, prostate, pancreatic, and breast cancer, as well as in leukemic cells [86–88]. Incorporation of sulforaphane in the diet was also shown to downregulate HDAC and to increase global histone acetylation, with specific increase at the *bax* and the *p21* promoter regions, in polyp tissue isolated from Apc^{min} mice and in PC-3 xenografts [89, 90]. Based on the structural similarities between the HDAC inhibitor trichostatin A and the sulforaphane metabolites sulforaphane-cysteine and sulforaphane-*N*-acetlcysteine revealed by molecular modeling, it was suggested, and then confirmed experimentally, that the metabolites were the actual inhibitors [86, 91]. Interestingly, targeted inhibition or knockdown of HDAC6 leads to acetylation of Hsp90 and disruption of its chaperone function, resulting in polyubiquitylation and depletion of Hsp90 client proteins, including BCR-ABL [92] and the androgen receptor [93]. Treatment with sulforaphane downregulates HDAC6's deacetylase activity, resulting in hyperacetylation of Hsp90 and inhibition of its association with the androgen receptor. Consequently, the proteasomal degradation of the androgen receptor is accelerated, leading to attenuation of androgen receptor-mediated signaling [93]. Thus, inhibition of Hsp90 activity by either disrupting its association with a co-chaperone or promoting its acetylation is a potential mechanism for HSF1 activation by sulforaphane.

6.5 The Importance of Reactivity with Sulfhydryl Groups for Activation of the KEAP1/NRF2/ARE Pathway and the Heat Shock Response

Reactivity with sulfhydryl groups is the only common feature of the phytochemicals discussed in this *Perspective*. Furthermore, the presence of this "chemical signature" is essential for activation of transcription factors NRF2 and HSF1. Thus, the central carbon of the isothiocyanate (—N=C=S) group of sulforaphane is highly electrophilic and reacts avidly with sulfhydryl groups. In addition, the isothiocyanates participate in transthiocarbamoylation reactions in which they are readily transferred from one (e.g., glutathione) conjugate to another sufhydryl group-containing molecule [94, 95]. Indeed, this reactivity with sulfhydryl groups underlies the cellular uptake and metabolism of sulforaphane, and the binding to its protein targets, including KEAP1 [96]. Similarly, the electrophilicity of the α,β-unsaturated carbonyl functionality within the chemical structures of celastrol, gedunin, withaferin A, and curcumin makes them highly reactive with sulfhydryl groups, although the epoxide moiety within the structures of gedunin and withaferin A may also contribute to this reactivity. In a high-throughput screen for inducers of the HSF1-dependent heat shock response that used a library of more than 80,000 compounds, the presence of the α,β-unsaturated carbonyl moiety was found to be required for both activation of the heat-shock response as well as for inhibition of glioma tumor cell growth [8]. In a recent large-scale study of approximately 1 million small molecules, electrophilicity was a prominent feature in several of the major clusters of more than 200 activators of the heat shock response that were identified [97]. The importance of sulfhydryl reactivity of the phytochemicals discussed here for activation of the KEAP1/NRF2/ARE pathway and the heat shock response is also supported by studies which have used these agents in combination with classical nucleophiles [5, 97, 98]. Thus, both induction of NRF2-dependent genes and of HSF1 target genes by celastrol are prevented by incubation with dithiothreitol (DTT) [5]. The ability of celastrol to increase the levels of Hsp70, and to decrease the levels of the Hsp90–Cdc37 complex can be reversed by *N*-acetylcysteine or glutathione, but not vitamin C, again implying that sulfhydryl reactivity is critical for these biological effects of celastrol [98]. Similarly, the celastrol-mediated degradation of the Hsp90 client protein ErbB2 is abolished by pretreatment of celastrol with DTT [43]. By use of one-dimensional (1D) and two-dimensional (2D) nuclear magnetic resonance (NMR) spectroscopy and application of density functional theory, it was recently shown that nucleophiles (e.g., cysteine and glutathione) add to celastrol regioselectively and stereospecifically to form Michael adducts, such that nucleophilic attack is favored exclusively at the α-face with the nucleophile approach *syn* to the α-C9 methyl [99]. Taken together, these findings imply that cysteine reactivity plays a critical role in triggering both the KEAP1/NRF2/ARE pathway and the heat shock response.

It is now widely recognized that cysteine residues within KEAP1 are the sensors for NRF2 activators. KEAP1 is a multidomain homodimeric protein which has five distinct domains (Fig. 6.4): NTR, N-terminal region (amino acids 1–60), broad complex, tramtrack, bric-á-brac (BTB; amino acids 61–179) which is a dimerization domain, IVR, intervening region (amino acids 180–314) which is a particularly cysteine-rich region containing eight cysteine residues among its 134 amino acids, Kelch domain (amino acids 315–598), through which KEAP1 binds to NRF2, and CTR, C-terminal region (amino acids 599–624). Murine KEAP1 contains 25 cysteine residues among its 624 amino acids (its human homolog has 27 cysteines), nine of which (i.e., C23, C38, C151, C241, C273, C288, C297, C319, and C613) are flanked by basic amino acids. The presence of neighboring basic amino acids is

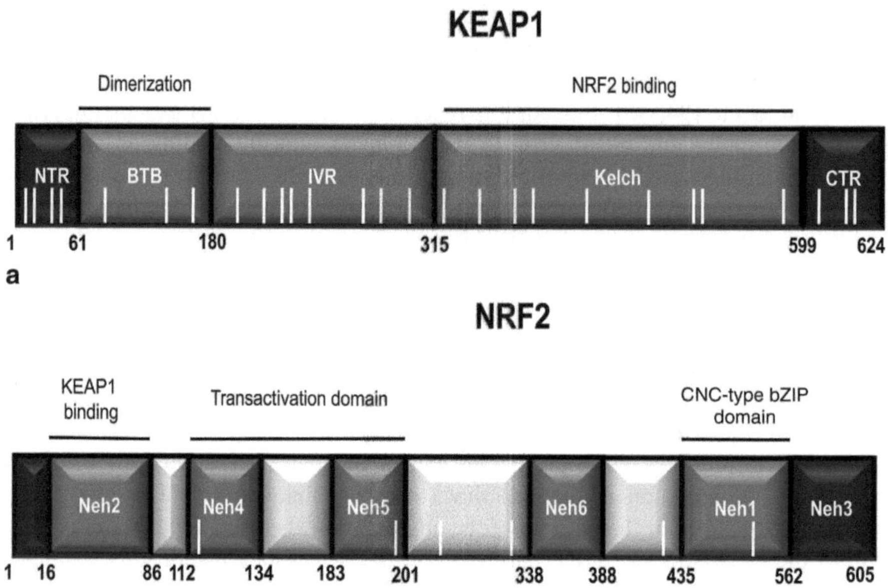

a

b

Fig. 6.4 Domain structure of KEAP1 (**a**) and NRF2 (**b**). In KEAP1, the BTB domain is the homodimerization domain and the site of interaction with Cullin 3. The Kelch domain is the NRF2-binding domain. In NRF2, the Neh2 domain is responsible for binding to KEAP1. Neh4 and Neh5 form the transactivation domain, and the Neh1 and Neh3 domains comprise the DNA-binding site of the transcription factor. The white bars indicate the distribution of the cysteine residues within the two proteins

known to lower the pK_a value of cysteine, favoring the formation of the thiolate anion at neutral pH, and thus increasing the cysteine reactivity [100]. Modifications of specific cysteine residues of KEAP1 by sulforaphane, or its sulfoxythiocarbamate derivative STCA, have been detected using both purified recombinant protein as well as lysates of cells that are expressing ectopically KEAP1 following exposure to these compounds [101–104]. By use of mass spectroscopy and mutagenesis analysis, it was established that C151 in the BTB domain, and C273 and C288 in the IVR domain are of particular importance for the repressor function of KEAP1, although depending on the reaction conditions and the experimental system, other cysteine residues can also be modified by sulforaphane, such as C38 in the N-terminal domain, C368, C489, and C583 in the Kelch domain, and C624 in the CTR [101, 104]. Mutation of C151 rendered KEAP1 a constitutive repressor of NRF2, which was unresponsive to induction by sulforaphane [10, 105]. In contrast, substitution of C273 or C288 with either serine or alanine led to complete loss of repressor activity and KEAP1 was unable to repress NRF2 even under basal conditions [105–107]. The increased activity of NRF2 in the presence of C273S/A or C288S/A mutant KEAP1 was caused by reduced ubiquitination of NRF2, resulting in stabilization of the transcription factor [10, 11]. Experiments using transgenic mice expressing either C273A or C288A KEAP1 mutants confirmed that these residues are required for repression of NRF2 under basal conditions [108].

Two studies—one conducted in zebrafish and another in cultured mammalian cells—have established that KEAP1 contains multiple inducer sensors. The study in zebrafish categorized a series of activators of NRF2 into two groups: those which react with C151 of KEAP1 (e.g., sulforaphane), and those which react with C273 (e.g., 15-deoxy-$\Delta^{12,14}$-prostaglandin J_2) [109]. When murine KEAP1 was ectopically expressed in mammalian cells, exposure of these cells to inducers of different types revealed that C151 and C288 each form discrete sensors, and there also exists a third sensor which is formed by H225, C226, and C613 [103]. Molecular modeling showed that C151 is in close spatial proximity with four positively charged amino acids, i.e., K131, R135, K150, and H154, an environment that most likely contributes to the increased reactivity of C151 by decreasing its pK_a and favoring the thiolate formation at physiological pH [103, 110]. Indeed, mutant KEAP1 in which K131, R135, and K150, were substituted with methionine residues had much lower sensitivity to sulforaphane [103]. Another model proposes that binding of inducers to C151 leads to a steric clash with residues in the adjacent α-helix, which may alter the interaction between KEAP1 and Cul3 [111]. It was recently suggested that gedunin may react directly with KEAP1 and impair its ability to target NRF2 for ubiquitination and proteasomal degradation by disrupting the association of KEAP1 with NRF2, although the cysteine reactivity of this phytochemical was not implicated in this model [48].

The importance of cysteine modifications for triggering the heat shock response is also apparent with regard to both HSF1 as well as its negative regulator Hsp90. Human and murine HSF1 contain five conserved cysteine residues. Based on the amino acid sequence of human HSF1, C153 and C373 might be predicted to be particularly reactive with electrophiles; they are in close proximity to basic amino acids (K148, K150, and K157 nearby C153; K372 directly adjacent to C373). An intermolecular disulfide bond formation between C36 and C103 within the human HSF1 has been reported to cause trimerization and DNA binding, whereas formation of an intramolecular disulfide bond (in which C153, C373, and C378 participate) inhibits trimerization and binding to heat shock elements in the promoter regions of heat shock genes [112]. Activation of murine HSF1 by H_2O_2 is dependent on C35 and C105, and the redox regulation of HSF1 is essential for induction of heat shock genes and for protection against thermal and oxidative stress [113]. C35 and C105 are localized within the DNA-binding domain of HSF1, and are required for disulfide bond formation in response to stress.

The reactivity of the cysteine residues of Hsp90 has also been investigated. C521 and C589/590 the rat Hsp90β, corresponding to C529 and C597/C598 in Hsp90α, have been predicted to be highly reactive [114]. Human Hsp90 is a dimeric multidomain protein that contains an N-terminal domain (amino acids 1–210) where ATP binds, a flexible linker region (amino acids 210–272) which affects the function, co-chaperone interaction, and conformation of Hsp90, a middle domain (amino acids 272–629) where many of its client proteins bind, and a C-terminal domain (amino acids 629–732) which contains a dimerization motif (Fig. 6.5) [115, 116].

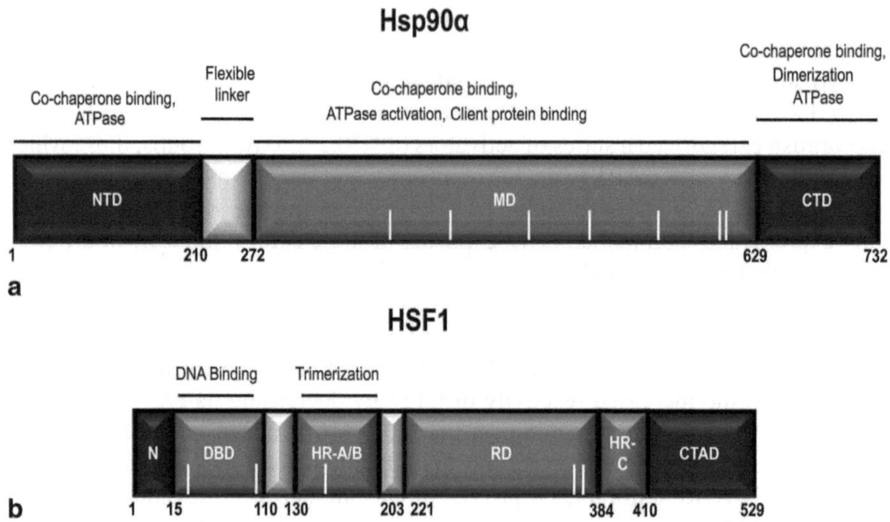

Fig. 6.5 Domain structure of human Hsp90α and HSF1. **a** In Hsp90α, the N-terminal domain (NTD) is the site of binding of ATP and some co-chaperones. The middle domain (MD) is where many of the Hsp90 client proteins and co-chaperones interact, and the C-terminal domain (CTD) contains a dimerization motif and a second ATP-binding site. **b** In HSF1, the DNA-binding domain (DBD) is at the N-terminus of the protein. Trimerization of the transcription factor is through the heptad repeat (HR-A/B) region and is negatively regulated by the HR-C domain. The transactivation domain (CTAD) is at the C-terminus of the protein. The regulatory domain (RD) has a negative regulatory function over the transactivation domain. The *white bars* indicate the distribution of the cysteine residues within the two proteins

In addition, the C-terminal domain has a conserved MEEVD amino acid sequence implicated in binding to co-chaperones with tetratricopeptide repeat (TRP) domains [117, 118], and can also bind to ATP when the N-terminal domain contains one ATP molecule. The binding of ATP causes the N-terminal domains to dimerize and become compacted, allowing them to function as a molecular clamp. Following ATP hydrolysis, the N-terminal domains dissociate. The ATPase cycle of Hsp90 is regulated at multiple levels. It can be stimulated by the co-chaperone Aha1 [119–121] or inhibited by Hop/Sti1 [122–124] and p23/Sba1 [125–129]. Posttranslational modifications of Hsp90 such as acetylation [130], phosphorylation [132, 132], and *S*-nitrosylation [133–137], represent another level of regulation. It has been shown that in human Hsp90α, *S*-nitrosylation at C597 inhibits the ATPase activity of the chaperone [137]. Further work by Retzlaff et al. [138] reported that substituting C597 in human Hsp90α with *S*-nitrosylation-mimicking residues, such as asparagine and aspartic acid, shifts the conformational equilibrium of the protein toward the open conformation, thus decreasing its chaperone activity. This conclusion is also supported by *in silico* studies indicating that C597 is involved in regulating the conformation in Hsp90 [139]. Whereas the identity of the individual cysteine residues of Hsp90 that could be modified by the phytochemicals discussed in this

Perspective is presently unknown, it is notable that C572, adjacent to a lysine residue, has been found to be modified by the electrophilic lipid peroxidation product 4-hydroxy-2-nonenal [140], whereas C521 was identified as the site of thiocarbamoylation when recombinant Hsp90α was incubated with 6-methylsulfinylhexyl isothiocyanate [95]. In agreement, by use of proteomics and click chemistry approaches, Hsp90 was identified as being modified when HEK293 cells were exposed to the sulfoxythiocarbamate derivative of sulforaphane, STCA [9, 102]. Because all cysteine residues reside in the middle client protein-binding domain of Hsp90, it is possible that their modification may lead to disruption of the Hsp90–client protein interactions.

Another possible target for sulfhydryl-reactive phytochemicals is the Hsp90 co-chaperone Cdc37. By use of heteronuclear single quantum coherence (HSQC) NMR spectroscopy, the interaction between the $^{1}H,^{15}N$-labeled N-terminal domain of Hsp90 with unlabeled full-length Cdc37 was investigated in the absence or presence of celastrol [141]. This approach, in combination with mutagenesis analysis and chemical modification (with *N*-ethylmaleimide) of the nine cysteine residues of Cdc37, revealed that celastrol is able to react with the three cysteine residues within the N-terminal domain of Cdc37. Reaction with celastrol leads to large conformational changes both in the N-terminal and in the middle Hsp90-binding domains of Cdc37, ultimately disrupting the Cdc37–Hsp90 association. As Cdc37 is an essential component of the Hsp90 complex machinery, its displacement from the complex will undoubtedly affect the function of the chaperone. In sum, the exact protein targets of the phytochemicals discussed here, which lead to activation of the heat shock response, are incompletely understood, likely to be multiple, and represent a subject of intense investigations by many laboratories. However, it is clear that the reactivity with cysteine residues within those protein targets is critical for the underlying mechanism(s).

It is noteworthy that activation of NRF2 generally occurs at lower inducer concentrations than those that activate HSF1. The most likely reason for this observation could be the relative nucleophilicity of the cysteine residues and their accessibility within the structure of the protein targets for reaction with the electrophilic inducers. Whereas the identification of the precise reason(s) requires further investigation, based on the available experimental evidence it could be proposed that, at low concentrations (Fig. 6.6a), inducers react first with KEAP1, activating the KEAP1/NRF2/ARE pathway. At higher inducer concentrations (Fig. 6.6b), a second target, such as Hsp90, or a co-chaperone within the Hsp90 complex machinery, is also affected, leading to induction of the heat shock response.

6.6 Conclusions

Celastrol, withaferin A, gedunin, curcumin, and sulforaphane are examples of structurally diverse phytochemicals with a common chemical signature: reactivity with sulfhydryl groups. This reactivity underlies their biological activities as

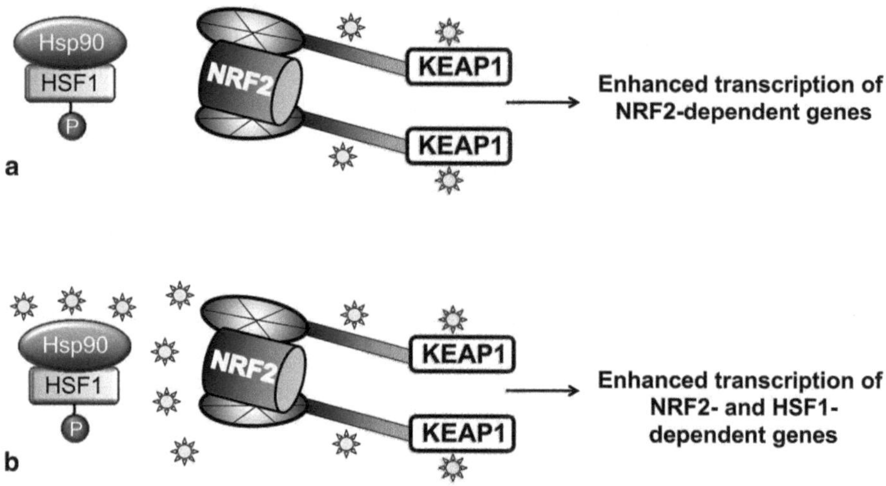

Fig. 6.6 Proposed model for concentration-dependent activation of the KEAP1/NRF2/ARE pathway and the heat shock response by sulfhydryl-reactive phytochemicals. **a** At low concentrations, inducers (depicted with solar symbols) react first with cysteine residues of KEAP1, activating transcription of NRF2-dependent genes. **b** At higher concentrations of inducers, a second target, such as Hsp90, HSF1, or a co-chaperone within the Hsp90 complex machinery, is also affected, leading to enhanced transcription of NRF2- and HSF1-dependent genes

multitarget agents for which protective effects have been documented in numerous animal models of human disease. The effects of such phytochemicals in biological systems are long lasting and comprehensive as they are due to induction of large networks of transcriptional programs regulated by transcription factors NRF2 and HSF1. The resulting upregulation of cytoprotective proteins provides the cell with multiple layers of protection against electrophiles, oxidants, and chronic inflammation, which are the major causes for the development of chronic degenerative conditions, such as cancer, cardiovascular disease, and neurodegeneration. Notably, lower concentrations of phytochemicals are required for induction of NRF2-dependent genes than those which induce HSF1-dependent responses, suggesting that activation of NRF2 precedes that of HSF1. It can be hypothesized that the KEAP1/NRF2/ARE pathway functions to defend against imminent danger. It is then followed by the heat shock response to protect against a potentially devastating damage and preserve the proteome. Collectively, the two pathways ensure adaptation and survival.

Acknowledgments The authors are extremely grateful to the Biotechnology and Biological Sciences Research Council (BBSRC, Project Grant BB/J007498/1) and Research Councils UK for financial support.

References

1. Talalay P, De Long MJ, Prochaska HJ (1988) Identification of a common chemical signal regulating the induction of enzymes that protect against chemical carcinogenesis. Proc Natl Acad Sci U S A 85:8261–8265
2. Talalay P, Fahey JW, Holtzclaw WD et al (1995) Chemoprotection against cancer by phase 2 enzyme induction. Toxicol Lett 82–83:173–179
3. Itoh K, Wakabayashi N, Katoh Y et al (1999) Keap1 represses nuclear activation of antioxidant responsive elements by Nrf2 through binding to the amino-terminal Neh2 domain. Genes Dev 13:76–86
4. Itoh K, Chiba T, Takahashi S et al (1997) An Nrf2/small Maf heterodimer mediates the induction of phase II detoxifying enzyme genes through antioxidant response elements. Biochem Biophys Res Commun 236:313–322
5. Trott A, West JD, Klaić L et al (2008) Activation of heat shock and antioxidant responses by the natural product celastrol: transcriptional signatures of a thiol-targeted molecule. Mol Biol Cell 19:1104–1112
6. Kansanen E, Jyrkkänen HK, Volger OL et al (2009) Nrf2-dependent and -independent responses to nitro-fatty acids in human endothelial cells: identification of heat shock response as the major pathway activated by nitro-oleic acid. J Biol Chem 284:33233–33241
7. Kansanen E, Jyrkkänen HK, Levonen AL (2012) Activation of stress signaling pathways by electrophilic oxidized and nitrated lipids. Free Radic Biol Med 52:973–982
8. Santagata S, Xu YM, Wijeratne EM et al (2012) Using the heat-shock response to discover anticancer compounds that target protein homeostasis. ACS Chem Biol 7:340–349
9. Zhang Y, Ahn YH, Benjamin IJ et al (2011) HSF1-dependent upregulation of Hsp70 by sulfhydryl-reactive inducers of the KEAP1/NRF2/ARE pathway. Chem Biol 18:1355–1361
10. Zhang DD, Lo SC, Cross JV et al (2004) Keap1 is a redox-regulated substrate adaptor protein for a Cul3-dependent ubiquitin ligase complex. Mol Cell Biol 24:10941–10953
11. Kobayashi A, Kang MI, Okawa H et al (2004) Oxidative stress sensor Keap1 functions as an adaptor for Cul3-based E3 ligase to regulate proteasomal degradation of Nrf2. Mol Cell Biol 24:7130–7139
12. Furukawa M, Xiong Y (2005) BTB protein Keap1 targets antioxidant transcription factor Nrf2 for ubiquitination by the Cullin 3-Roc1 ligase. Mol Cell Biol 25:162–171
13. Rada P, Rojo AI, Chowdhry S et al (2011) SCF/β-TrCP promotes glycogen synthase kinase 3-dependent degradation of the Nrf2 transcription factor in a Keap1-independent manner. Mol Cell Biol 31:1121–1133
14. Baird L, Dinkova-Kostova AT (2011) The cytoprotective role of the Keap1-Nrf2 pathway. Arch Toxicol 85:241–272
15. Dinkova-Kostova AT, Holtzclaw WD, Kensler TW (2005) The role of Keap1 in cellular protective responses. Chem Res Toxicol 18:1779–1791
16. Kobayashi M, Yamamoto M (2006) Nrf2-Keap1 regulation of cellular defense mechanisms against electrophiles and reactive oxygen species. Adv Enzyme Regul 46:113–140
17. Kensler TW, Wakabayashi N, Biswal S (2007) Cell survival responses to environmental stresses via the Keap1-Nrf2-ARE pathway. Annu Rev Pharmacol Toxicol 47:89–116
18. Malhotra D, Portales-Casamar E, Singh A et al (2010) Global mapping of binding sites for Nrf2 identifies novel targets in cell survival response through ChIP-Seq profiling and network analysis. Nucleic Acids Res 38:5718–5734
19. Akerfelt M, Morimoto RI, Sistonen L (2010) Heat shock factors: integrators of cell stress, development and lifespan. Nat Rev Mol Cell Biol 11:545–555
20. Morimoto RI (2011) The Heat Shock Response: Systems Biology of Proteotoxic Stress in Aging and Disease. Cold Spring Harb Symp Quant Biol 76:91–99
21. Morimoto RI, Santoro MG (1998) Stress-inducible responses and heat shock proteins: new pharmacologic targets for cytoprotection. Nat Biotechnol 16:833–838

22. Westerheide SD, Morimoto RI (2005) Heat shock response modulators as therapeutic tools for diseases of protein conformation. J Biol Chem 280:33097–33100
23. Anckar J, Sistonen L (2011) Regulation of HSF1 function in the heat stress response: implications in aging and disease. Annu Rev Biochem 80:1089–1115
24. Zou J, Guo Y, Guettouche T et al (1998) Repression of heat shock transcription factor HSF1 activation by HSP90 (HSP90 complex) that forms a stress-sensitive complex with HSF1. Cell 94:471–480
25. Ali A, Bharadwaj S, O'Carroll R et al (1998) HSP90 interacts with and regulates the activity of heat shock factor 1 in Xenopus oocytes. Mol Cell Biol 18:4949–4960
26. Richter K, Haslbeck M, Buchner J (2010) The heat shock response: life on the verge of death. Mol Cell 40:253–266
27. Westerheide SD, Bosman JD, Mbadugha BN et al (2004) Celastrols as inducers of the heat shock response and cytoprotection. J Biol Chem 279:56053–56060
28. Hieronymus H, Lamb J, Ross KN et al (2006) Gene expression signature-based chemical genomic prediction identifies a novel class of HSP90 pathway modulators. Cancer Cell 10:321–330
29. Lamb J, Crawford ED, Peck D et al (2006) The connectivity map: using gene-expression signatures to connect small molecules, genes, and disease. Science 313:1929–1935
30. Matts RL, Brandt GE, Lu Y et al (2011) A systematic protocol for the characterization of Hsp90 modulators. Bioorg Med Chem 19:684–692
31. Chadli A, Felts SJ, Wang Q et al (2010) Celastrol inhibits Hsp90 chaperoning of steroid receptors by inducing fibrillization of the co-chaperone p23. J Biol Chem 285:4224–4231
32. Zhang T, Hamza A, Cao X et al (2008) A novel Hsp90 inhibitor to disrupt Hsp90/Cdc37 complex against pancreatic cancer cells. Mol Cancer Ther 7:162–170
33. Zhang T, Li Y, Yu Y et al (2009) Characterization of celastrol to inhibit hsp90 and cdc37 interaction. J Biol Chem 284:35381–35389
34. Hansen J, Palmfeldt J, Vang S et al (2011) Quantitative proteomics reveals cellular targets of celastrol. PLoS One 6:e26634
35. Wu F, Han M, Wilson JX (2009) Tripterine prevents endothelial barrier dysfunction by inhibiting endogenous peroxynitrite formation. Br J Pharmacol 157:1014–1023
36. Francis SP, Kramarenko II, Brandon CS et al (2011) Celastrol inhibits aminoglycoside-induced ototoxicity via heat shock protein 32. Cell Death Dis 2:e195
37. Zhang YQ, Sarge KD (2007) Celastrol inhibits polyglutamine aggregation and toxicity though induction of the heat shock response. J Mol Med (Berl) 85:1421–1428
38. Paris D, Ganey NJ, Laporte V et al (2010) Reduction of beta-amyloid pathology by celastrol in a transgenic mouse model of Alzheimer's disease. J Neuroinflammation 7:17
39. Allison AC, Cacabelos R, Lombardi VR et al (2001) Celastrol, a potent antioxidant and anti-inflammatory drug, as a possible treatment for Alzheimer's disease. Prog Neuropsychopharmacol Biol Psychiatry 25:1341–1357
40. Cleren C, Calingasan NY, Chen J et al (2005) Celastrol protects against MPTP- and 3-nitropropionic acid-induced neurotoxicity. J Neurochem 94:995–1004
41. Kiaei M, Kipiani K, Petri S et al (2005) Celastrol blocks neuronal cell death and extends life in transgenic mouse model of amyotrophic lateral sclerosis. Neurodegener Dis 2:246–254
42. Hughes D, Guttenplan JB, Marcus CB et al (2008) Heat shock protein 90 inhibitors suppress aryl hydrocarbon receptor-mediated activation of CYP1A1 and CYP1B1 transcription and DNA adduct formation. Cancer Prev Res (Phila) 1:485–493
43. Raja SM, Clubb RJ, Ortega-Cava C et al (2011) Anticancer activity of Celastrol in combination with ErbB2-targeted therapeutics for treatment of ErbB2-overexpressing breast cancers. Cancer Biol Ther 11:263–276
44. Yadav VR, Sung B, Prasad S et al (2010) Celastrol suppresses invasion of colon and pancreatic cancer cells through the downregulation of expression of CXCR4 chemokine receptor. J Mol Med (Berl) 88:1243–1253
45. Huang L, Zhang Z, Zhang S et al (2011) Inhibitory action of celastrol on hypoxia-mediated angiogenesis and metastasis via the HIF-1α pathway. Int J Mol Med 27:407–415

46. Lee JH, Choi KJ, Seo WD et al (2011) Enhancement of radiation sensitivity in lung cancer cells by celastrol is mediated by inhibition of Hsp90. Int J Mol Med 27:441–446
47. Brandt GE, Schmidt MD, Prisinzano TE et al (2008) Gedunin, a novel Hsp90 inhibitor: semisynthesis of derivatives and preliminary structure-activity relationships. J Med Chem 51:6495–6502
48. Smirnova NA, Haskew-Layton RE, Basso M et al (2011) Development of Neh2-luciferase reporter and its application for high throughput screening and real-time monitoring of Nrf2 activators. Chem Biol 2011 18:752–765
49. Uddin SJ, Nahar L, Shilpi JA et al (2007) Gedunin, a limonoid from Xylocarpus granatum, inhibits the growth of CaCo-2 colon cancer cell line in vitro. Phytother Res 21:757–761
50. Cazal CM, Choosang K, Severino VG et al (2010) Evaluation of effect of triterpenes and limonoids on cell growth, cell cycle and apoptosis in human tumor cell line. Anticancer Agents Med Chem 10:769–776
51. Kamath SG, Chen N, Xiong Y et al (2009) Gedunin, a novel natural substance, inhibits ovarian cancer cell proliferation. Int J Gynecol Cancer 19:1564–1569
52. Zhang B, Au Q, Yoon IS et al (2009) Identification of small-molecule HSF1 amplifiers by high content screening in protection of cells from stress induced injury. V Biochem Biophys Res Commun 390:925–930
53. Su BN, Park EJ, Nikolic D et al (2003) Isolation and characterization of miscellaneous secondary metabolites of Deprea subtriflora. J Nat Prod 66:1089–1093
54. Kang YH, Pezzuto JM (2004) Induction of quinone reductase as a primary screen for natural product anticarcinogens. Methods Enzymol 382:380–414
55. Yu Y, Hamza A, Zhang T et al (2010) Withaferin A targets heat shock protein 90 in pancreatic cancer cells. Biochem Pharmacol 79:542–551
56. Grover A, Shandilya A, Agrawal V et al (2011) Blocking the chaperone kinome pathway: mechanistic insights into a novel dual inhibition approach for supra-additive suppression of malignant tumors. Biochem Biophys Res Commun 404:498–503
57. Grover A, Shandilya A, Agrawal V et al (2011) Hsp90/Cdc37 chaperone/co-chaperone complex, a novel junction anticancer target elucidated by the mode of action of herbal drug Withaferin A. BMC Bioinformatics 12(Suppl 1):S30
58. Zhang X, Mukerji R, Samadi AK et al (2011) Down-regulation of estrogen receptor-alpha and rearranged during transfection tyrosine kinase is associated with withaferin A-induced apoptosis in MCF-7 breast cancer cells. BMC Complement Altern Med 11:84
59. Dinkova-Kostova AT, Talalay P (1999) Relation of structure of curcumin analogs to their potencies as inducers of Phase 2 detoxification enzymes. Carcinogenesis 20:911–914
60. Dinkova-Kostova AT, Massiah MA, Bozak RE et al (2001) Potency of Michael reaction acceptors as inducers of enzymes that protect against carcinogenesis depends on their reactivity with sulfhydryl groups. Proc Natl Acad Sci U S A 98:3404–3409
61. Scapagnini G, Colombrita C, Amadio M et al (2006) Curcumin activates defensive genes and protects neurons against oxidative stress. Antioxid Redox Signal 8:395–403
62. Balogun E, Hoque M, Gong P et al (2003) Curcumin activates the haem oxygenase-1 gene via regulation of Nrf2 and the antioxidant-responsive element. Biochem J 371:887–895
63. Jeong GS, Oh GS, Pae HO et al (2006) Comparative effects of curcuminoids on endothelial heme oxygenase-1 expression: ortho-methoxy groups are essential to enhance heme oxygenase activity and protection. Exp Mol Med 38:393–400
64. McNally SJ, Harrison EM, Ross JA et al (2006) Curcumin induces heme oxygenase-1 in hepatocytes and is protective in simulated cold preservation and warm reperfusion injury. Transplantation 81:623–626
65. Rushworth SA, Ogborne RM, Charalambos CA et al (2006) Role of protein kinase C delta in curcumin-induced antioxidant response element-mediated gene expression in human monocytes. Biochem Biophys Res Commun 341:1007–1016
66. Andreadi CK, Howells LM, Atherfold PA et al (2006) Involvement of Nrf2, p38, B-Raf, and nuclear factor-kappaB, but not phosphatidylinositol 3-kinase, in induction of hemeoxygenase-1 by dietary polyphenols. Mol Pharmacol 69:1033–1040

67. Sood A, Mathew R, Trachtman H (2001) Cytoprotective effect of curcumin in human proximal tubule epithelial cells exposed to shiga toxin. Biochem Biophys Res Commun 283:36–41
68. Kanitkar M, Bhonde RR (2008) Curcumin treatment enhances islet recovery by induction of heat shock response proteins, Hsp70 and heme oxygenase-1, during cryopreservation. Life Sci 82:182–189
69. Khan S, Heikkila JJ (2011) Curcumin-induced inhibition of proteasomal activity, enhanced HSP accumulation and the acquisition of thermotolerance in Xenopus laevis A6 cells. Comp Biochem Physiol A Mol Integr Physiol 158:566–576
70. Wu LX, Xu JH, Huang XW et al (2006) Down-regulation of p210(bcr/abl) by curcumin involves disrupting molecular chaperone functions of Hsp90. Acta Pharmacol Sin 27:694–699
71. Chen HW, Yu SL, Chen JJ et al (2004) Anti-invasive gene expression profile of curcumin in lung adenocarcinoma based on a high throughput microarray analysis. Mol Pharmacol 65:99–110
72. Shen SQ, Zhang Y, Xiang JJ et al (2007) Protective effect of curcumin against liver warm ischemia/reperfusion injury in rat model is associated with regulation of heat shock protein and antioxidant enzymes. World J Gastroenterol 13:1953–1961
73. Kato K, Ito H, Kamei K et al (1998) Stimulation of the stress-induced expression of stress proteins by curcumin in cultured cells and in rat tissues in vivo. Cell Stress Chaperones 3:152–160
74. Surh YJ, Chun KS (2007) Cancer chemopreventive effects of curcumin. Adv Exp Med Biol 595:149–172
75. Calabrese V, Bates TE, Mancuso C et al (2008) Curcumin and the cellular stress response in free radical-related diseases. Mol Nutr Food Res 52:1062–1073
76. Hatcher H, Planalp R, Cho J et al (2008) Curcumin: from ancient medicine to current clinical trials. Cell Mol Life Sci 65:1631–1652
77. Lao CD, Ruffin MT 4th, Normolle D et al (2006) Dose escalation of a curcuminoid formulation. BMC Complement Altern Med 6:10
78. Zhang Y, Talalay P, Cho CG et al (1992) A major inducer of anticarcinogenic protective enzymes from broccoli: isolation and elucidation of structure. Proc Natl Acad Sci U S A 89:2399–2403
79. Dinkova-Kostova AT, Talalay P (2008) Direct and indirect antioxidant properties of inducers of cytoprotective proteins. Mol Nutr Food Res 52(Suppl 1):S128–S138
80. Hayes JD, McMahon M, Chowdhry S et al (2010) Cancer chemoprevention mechanisms mediated through the Keap1-Nrf2 pathway. Antioxid Redox Signal 13:1713–1748
81. Calabrese V, Cornelius C, Dinkova-Kostova AT et al (2010) Cellular stress responses, the hormesis paradigm, and vitagenes: novel targets for therapeutic intervention in neurodegenerative disorders. Antioxid Redox Signal 13:1763–1811
82. Hu R, Xu C, Shen G et al (2006) Gene expression profiles induced by cancer chemopreventive isothiocyanate sulforaphane in the liver of C57BL/6J mice and C57BL/6J/Nrf2 (-/-) mice. Cancer Lett 243:170–192
83. Sharma R, Sharma A, Chaudhary P et al (2010) Role of lipid peroxidation in cellular responses to D, L-sulforaphane, a promising cancer chemopreventive agent. Biochemistry 49:3191–3202
84. Gan N, Wu YC, Brunet M et al (2010) Sulforaphane activates heat shock response and enhances proteasome activity through up-regulation of Hsp27. J Biol Chem 285:35528–35536
85. Li Y, Zhang T, Schwartz SJ et al (2011) Sulforaphane potentiates the efficacy of 17-allylamino 17-demethoxygeldanamycin against pancreatic cancer through enhanced abrogation of Hsp90 chaperone function. Nutr Cancer 63:1151–1159
86. Myzak MC, Karplus PA, Chung FL et al (2004) A novel mechanism of chemoprotection by sulforaphane: inhibition of histone deacetylase. Cancer Res 64:5767–5774
87. Myzak MC, Hardin K, Wang R et al (2006) Sulforaphane inhibits histone deacetylase activity in BPH-1, LnCaP and PC-3 prostate epithelial cells. Carcinogenesis 27:811–819
88. Pledgie-Tracy A, Sobolewski MD, Davidson NE (2007) Sulforaphane induces cell type-specific apoptosis in human breast cancer cell lines. Mol Cancer Ther 6:1013–1021
89. Myzak MC, Dashwood WM, Orner GA et al (2006) Sulforaphane inhibits histone deacetylase in vivo and suppresses tumorigenesis in Apc-minus mice. FASEB J 20:506–508

90. Myzak MC, Tong P, Dashwood WM et al (2007) Sulforaphane retards the growth of human PC-3 xenografts and inhibits HDAC activity in human subjects. Exp Biol Med (Maywood) 232:227–234
91. Dashwood RH, Ho E (2007) Dietary histone deacetylase inhibitors: from cells to mice to man. Semin Cancer Biol 17:363–369
92. Bali P, Pranpat M, Bradner J et al (2005) Inhibition of histone deacetylase 6 acetylates and disrupts the chaperone function of heat shock protein 90: a novel basis for antileukemia activity of histone deacetylase inhibitors. J Biol Chem 280:26729–26734
93. Gibbs A, Schwartzman J, Deng V et al (2009) Sulforaphane destabilizes the androgen receptor in prostate cancer cells by inactivating histone deacetylase 6. Proc Natl Acad Sci U S A 106:16663–16668
94. Baillie TA, Kassahun K (1994) Reversibility in glutathione-conjugate formation. Adv Pharmacol 27:163–181
95. Shibata T, Kimura Y, Mukai A et al (2011) Transthiocarbamoylation of proteins by thiolated isothiocyanates. J Biol Chem 286:42150–42161
96. Zhang Y (2012) The molecular basis that unifies the metabolism, cellular uptake and chemopreventive activities of dietary isothiocyanates. Carcinogenesis 33:2–9
97. Calamini B, Silva MC, Madoux F et al (2011) Small-molecule proteostasis regulators for protein conformational diseases. Nat Chem Biol 8:185–196
98. Peng B, Xu L, Cao F et al (2010) HSP90 inhibitor, celastrol, arrests human monocytic leukemia cell U937 at G0/G1 in thiol-containing agents reversible way. Mol Cancer 9:79
99. Klaić L, Trippier PC, Mishra RK et al (2011) Remarkable stereospecific conjugate additions to the Hsp90 inhibitor celastrol. J Am Chem Soc 133:19634–19637
100. Snyder GH, Cennerazzo MJ, Karalis AJ et al (1981) Electrostatic influence of local cysteine environments on disulfide exchange kinetics. Biochemistry 20:6509–6519
101. Hong F, Freeman ML, Liebler DC (2005) Identification of sensor cysteines in human Keap1 modified by the cancer chemopreventive agent sulforaphane. Chem Res Toxicol 18:1917–1926
102. Ahn YH, Hwang Y, Liu H et al (2010) Electrophilic tuning of the chemoprotective natural product sulforaphane. Proc Natl Acad Sci U S A 107:9590–9595
103. McMahon M, Lamont DJ, Beattie KA et al (2010) Keap1 perceives stress via three sensors for the endogenous signaling molecules nitric oxide, zinc, and alkenals. Proc Natl Acad Sci U S A 107:18838–18843
104. Hu C, Eggler AL, Mesecar AD et al (2011) Modification of Keap1 cysteine residues by sulforaphane. Chem Res Toxicol 24:515–521
105. Zhang DD, Hannink M (2003) Distinct cysteine residues in Keap1 are required for Keap1-dependent ubiquitination of Nrf2 and for stabilization of Nrf2 by chemopreventive agents and oxidative stress. Mol Cell Biol 23:8137–8151
106. Levonen AL, Landar A, Ramachandran A et al (2004) Cellular mechanisms of redox cell signalling: role of cysteine modification in controlling antioxidant defences in response to electrophilic lipid oxidation products. Biochem J 378:373–382
107. Wakabayashi N, Dinkova-Kostova AT, Holtzclaw WD et al (2004) Protection against electrophile and oxidant stress by induction of the phase 2 response: fate of cysteines of the Keap1 sensor modified by inducers. Proc Natl Acad Sci U S A 101:2040–2045
108. Yamamoto T, Suzuki T, Kobayashi A et al (2008) Physiological significance of reactive cysteine residues of Keap1 in determining Nrf2 activity. Mol Cell Biol 28:2758–2770
109. Kobayashi M, Li L, Iwamoto N et al (2009) The antioxidant defense system Keap1-Nrf2 comprises a multiple sensing mechanism for responding to a wide range of chemical compounds. Mol Cell Biol 29:493–502
110. Fourquet S, Guerois R, Biard D et al (2010) Activation of NRF2 by nitrosative agents and H2O2 involves KEAP1 disulfide formation. J Biol Chem 285:8463–8471
111. Eggler AL, Small E, Hannink M et al (2009) Cul3-mediated Nrf2 ubiquitination and antioxidant response element (ARE) activation are dependent on the partial molar volume at position 151 of Keap1. Biochem J 422:171–180

112. Lu M, Kim HE, Li CR et al (2008) Two distinct disulfide bonds formed in human heat shock transcription factor 1 act in opposition to regulate its DNA binding activity. Biochemistry 47:6007–6015
113. Ahn SG, Thiele DJ (2003) Redox regulation of mammalian heat shock factor 1 is essential for Hsp gene activation and protection from stress. Genes Dev 17:516–528
114. Nardai G, Sass B, Eber J et al (2000) Reactive cysteines of the 90-kDa heat shock protein, Hsp90. Arch Biochem Biophys 384:59–67
115. Whitesell L, Lindquist SL (2005) HSP90 and the chaperoning of cancer. Nat Rev Cancer 5:761–772
116. Tsutsumi S, Mollapour M, Prodromou C et al (2012) Charged linker sequence modulates eukaryotic heat shock protein 90 (Hsp90) chaperone activity. Proc Natl Acad Sci U S A 109:2937–2942
117. Chen S, Sullivan WP, Toft DO et al (1998) Differential interactions of p23 and the TPR-containing proteins Hop, Cyp40, FKBP52 and FKBP51 with Hsp90 mutants. Cell Stress Chaperones 3:118–129
118. Young JC, Obermann WM, Hartl FU (1998) Specific binding of tetratricopeptide repeat proteins to the C-terminal 12-kDa domain of Hsp90. J Biol Chem 273:18007–18010
119. Meyer P, Prodromou C, Liao C et al (2004) Structural basis for recruitment of the ATPase activator Aha1 to the Hsp90 chaperone machinery. EMBO J 23:1402–1410
120. Panaretou B, Siligardi G, Meyer P et al (2002) Activation of the ATPase activity of hsp90 by the stress-regulated cochaperone Aha1. Mol Cell 10:1307–1318
121. Retzlaff M, Hagn F, Mitschke L et al (2010) Asymmetric activation of the Hsp90 dimer by its cochaperone Aha1. Mol Cell 37:344–354
122. Prodromou C, Siligardi G, O'Brien R et al (1999) Regulation of Hsp90 ATPase activity by tetratricopeptide repeat (TPR)-domain co-chaperones. EMBO J 18:754–762
123. Richter K, Muschler P, Hainzl O et al (2003) Sti1 is a non-competitive inhibitor of the Hsp90 ATPase. Binding prevents the N-terminal dimerization reaction during the ATPase cycle. J Biol Chem 278:10328–10333
124. Lee CT, Graf C, Mayer FJ et al (2012) Dynamics of the regulation of Hsp90 by the co-chaperone Sti1. EMBO J 31:1518–1528
125. Forafonov F, Toogun OA, Grad I et al (2008) p23/Sba1p protects against Hsp90 inhibitors independently of its intrinsic chaperone activity. Mol Cell Biol 28:3446–3456
126. Johnson JL, Beito TG, Krco CJ et al (1994) Characterization of a novel 23-kilodalton protein of unactive progesterone receptor complexes. Mol Cell Biol 14:1956–1963
127. Richter K, Walter S, Buchner J (2004) The Co-chaperone Sba1 connects the ATPase reaction of Hsp90 to the progression of the chaperone cycle. J Mol Biol 342:1403–1413
128. Prodromou C, Panaretou B, Chohan S et al (2000) The ATPase cycle of Hsp90 drives a molecular 'clamp' via transient dimerization of the N-terminal domains. EMBO J 19:4383–4392
129. Young JC, Hartl FU (2000) Polypeptide release by Hsp90 involves ATP hydrolysis and is enhanced by the co-chaperone p23. EMBO J 19:5930–5940
130. Scroggins BT, Robzyk K, Wang D et al (2007) An acetylation site in the middle domain of Hsp90 regulates chaperone function. Mol Cell 25:151–159
131. Wandinger SK, Suhre MH, Wegele H et al (2006) The phosphatase Ppt1 is a dedicated regulator of the molecular chaperone Hsp90. EMBO J 25:367–376
132. Mollapour M, Tsutsumi S, Donnelly AC et al (2010) Swe1Wee1-dependent tyrosine phosphorylation of Hsp90 regulates distinct facets of chaperone function. Mol Cell 37:333–343
133. Jorge I, Casas EM, Villar M et al (2007) High-sensitivity analysis of specific peptides in complex samples by selected MS/MS ion monitoring and linear ion trap mass spectrometry: application to biological studies. J Mass Spectrom 42:1391–1403
134. Rhee KY, Erdjument-Bromage H, Tempst P et al (2005) S-nitroso proteome of Mycobacterium tuberculosis: Enzymes of intermediary metabolism and antioxidant defense. Proc Natl Acad Sci U S A 102:467–472
135. Lindermayr C, Saalbach G, Durner J (2005) Proteomic identification of S-nitrosylated proteins in Arabidopsis. Plant Physiol 137:921–930

136. Zhang Y, Keszler A, Broniowska KA et al (2005) Characterization and application of the biotin-switch assay for the identification of S-nitrosated proteins. Free Radic Biol Med 38:874–881
137. Martínez-Ruiz A, Villanueva L, González de Orduña C et al (2005) S-nitrosylation of Hsp90 promotes the inhibition of its ATPase and endothelial nitric oxide synthase regulatory activities. Proc Natl Acad Sci U S A 102:8525–8530
138. Retzlaff M, Stahl M, Eberl HC et al (2009) Hsp90 is regulated by a switch point in the C-terminal domain. EMBO Rep 10:1147–1153
139. Morra G, Verkhivker G, Colombo G (2009) Modeling signal propagation mechanisms and ligand-based conformational dynamics of the Hsp90 molecular chaperone full-length dimer. PLoS Comput Biol 5:e1000323
140. Carbone DL, Doorn JA, Kiebler Z et al (2005) Modification of heat shock protein 90 by 4-hydroxynonenal in a rat model of chronic alcoholic liver disease. J Pharmacol Exp Ther 315:8–15
141. Sreeramulu S, Gande SL, Göbel M et al (2009) Molecular mechanism of inhibition of the human protein complex Hsp90-Cdc37, a kinome chaperone-cochaperone, by triterpene celastrol. Angew Chem Int Ed Engl 48:5853–5855

Chapter 7
Chemical and Biological Mechanisms of Phytochemical Activation of NRF2 and Importance in Disease Prevention

Aimee L. Eggler and Sergey N. Savinov

Abstract Plants are an incredibly rich source of compounds that activate the Nrf2 transcription factor, leading to upregulation of a battery of cytoprotective genes. This perspective surveys established and proposed molecular mechanisms of Nrf2 activation by phytochemicals with a special emphasis on a common chemical property of Nrf2 activators: the ability as "soft" electrophiles to modify cellular thiols, either directly or as oxidized biotransformants. In addition, the role of reactive oxygen/nitrogen species as secondary messengers in Nrf2 activation is discussed. While the uniquely reactive C151 of Keap1, an Nrf2 repressor protein, is highlighted as a key target of cytoprotective phytochemicals, also reviewed are other stress-responsive proteins, including kinases, which play nonredundant roles in the activation of Nrf2 by plant-derived agents. Finally, the Perspective presents two key factors accounting for the enhanced therapeutic windows of effective phytochemical activators of the Keap1–Nrf2 axis: enhanced selectivity toward sensor cysteines and reversibility of addition to thiolate molecules.

Abbreviations

Nrf2	NF-E2-related factor 2
ARE	antioxidant response element
ROS/RNS	reactive oxygen/nitrogen species
QR1	NAD(P)H: quinone oxidoreductase 1
wt	wild-type
GSH	glutathione
NAC	*N*-acetylcysteine
EGCG	epigallocatechin gallate

A. L. Eggler (✉)
Department of Chemistry, Villanova University, Villanova, PA 19085, USA
e-mail: aimee.eggler@villanova.edu

S. N. Savinov
Purdue University Center for Cancer Research, West Lafayette, IN 47907, USA

D. R. Gang (ed.), *50 Years of Phytochemistry Research,*
Recent Advances in Phytochemistry 43, DOI 10.1007/978-3-319-00581-2_7,

tBHQ	*tert*-butylhydroquinone
DAS	diallylsulfide
CYP2E1	cytochrome P450 2E1
DASO	diallyl sulfoxide
$DASO_2$	diallyl sulfone
I3C	indole-3-carbinol
DIM	3,3′-diindolylmethane
3MI	3-methylindole
CYP	cytochrome P450
AhR	aryl hydrocarbon receptor
β-NF	β-naphthoflavone
ERβ	estrogen receptor β
QR2	quinone reductase 2
SOD	superoxide dismutase
HO-1	heme oxygenase 1
BAECs	bovine aortic endothelial cells
GSTs	glutathione *S*-transferases
MRPs	multidrug resistance-associated proteins
PI3K	phosphatidylinositol 3-kinase
MAPKs	mitogen-activated protein kinases
PEITC	phenethyl isothiocyanate
PKC	protein kinase C

7.1 Introduction

Numerous phytochemicals have shown great promise for prevention and treatment of various human diseases. For example, ClinicalTrials.gov lists 25 different intervention trials investigating the effects of standardized preparations of broccoli sprouts. These trials include the amelioration of symptoms of diseases as diverse as autism,[1] cystic fibrosis, influenza, asthma, and chronic obstructive pulmonary disease. The trials also encompass studies on the prevention of breast, lung, and prostate cancers, carcinogenesis from aflatoxin exposure, and cardiovascular disease. The purpose of this Perspective is to (1) review a key molecular mechanism of phytochemicals in the prevention and amelioration of diseases: the activation of the transcription factor NF-E2-related factor 2 (Nrf2); (2) outline common chemical features of potent Nrf2 activators; and (3) provide perspectives for harnessing these features for more effective disease prevention or treatment.

[1] ClinicalTrials.gov identifiers, in order as listed above—amelioration studies: NCT01474993, NCT01315665, NCT01269723, NCT01183923, and NCT00994604; prevention studies: NCT00982319, NCT00255775, NCT00607932, NCT01437501, and NCT00252018.

7.2 Upregulation of Cytoprotective Genes by Nrf2

Oxidative stress and associated inflammation contribute to the progression of many chronic degenerative diseases in humans [1, 2]. Exposure to oxidant and electrophilic agents from air, water, food, and other environmental sources has also been implicated in a large (70–90 %) component of cancer and cardiovascular disease risks [3]. The Nrf2 transcription factor has emerged as a key player in protecting cells against various intrinsic and extrinsic assaults. Nrf2 regulates more than 600 genes, including over 100 that encode cytoprotective proteins [4], named for their ability to protect cells against oxidative stress, reactive electrophilic species, and other types of stress (reviewed in [5]). In brief, these proteins include antioxidant enzymes, NADPH regeneration enzymes, glutathione (GSH) biosynthesis enzymes, heat shock proteins, enzymes that facilitate the elimination of xenobiotic toxicants such as detoxification enzymes and drug-efflux pumps, as well as subunits of the 26S proteasome. Collectively, these Nrf2-regulated genes share at least one copy of the antioxidant response element (ARE, $5'\text{-}{}^{A}/_{G}TGA{}^{C}/_{G}{}^{T}/_{A}{}^{C}/_{N}{}^{A}/_{N}GC{}^{A}/_{T}\text{-}3'$) in their promoter region [4]. Nrf2 binds to the ARE as a heterodimer with one of several small Maf transcription factors, leading to upregulation of gene transcription.

Upregulation of this battery of cytoprotective proteins through Nrf2 activation is a critical component of an organism's ability to cope with intrinsic and extrinsic stress factors, including inflammation, reactive oxygen/nitrogen species (ROS/RNS), shear stressors such as endothelial stressors, and environmental toxins. Studies on *Nrf2*$^{-/-}$ mice, which have low and largely noninducible levels of many cytoprotective proteins [4], serve as striking and comprehensive examples of the importance of these proteins in maintaining health and preventing disease. Nrf2-deficient mice are prone to develop disorders that are caused by ROS and inflammation, including macular degeneration [6], neurodegeneration in a murine model of Parkinson's disease [7], cardiac disorders [8, 9], and chemically induced tumorigenesis [10–12]. Furthermore, *Nrf2*$^{-/-}$ mice are more susceptible to damage of the blood–brain barrier following brain injury [13], formation of carcinogen–DNA adducts in the lung exposed to diesel exhaust [14], acute pulmonary injury induced by butylated hydroxytoluene [15], and hepatic damage by acetaminophen [16]. In addition, they are deficient in their intrinsic capacity for skin wound healing [17]. Since loss of Nrf2 increases pathological cell and tissue damage in response to intrinsic and extrinsic factors, it is believed that upregulation of Nrf2 serves both cytoprotective and preventive roles in diverse pathophysiological situations.

In support of the importance of Nrf2 in preventing human diseases, inherited DNA polymorphisms that reduce the abundance of Nrf2 are associated with various pathologies, including chronic gastritis, ulcerative colitis, skin pathologies such as skin vitiligo, and adult respiratory distress syndrome (reviewed in [18]). In fact, a number of clinical trials that assess the effects of broccoli sprouts include direct measurements of Nrf2 activation. For example, Nrf2 levels, along with markers of oxidative stress, will be assessed after administration of macerated broccoli sprouts in both healthy volunteers and those with cystic fibrosis,[2] in nasal epithelial cells obtained by curettage, as well as in alveolar macrophages and bronchial epithelial cells of patients with chronic obstructive pulmonary disease.[3]

7.3 Phytochemical Activation of Nrf2

7.3.1 Overview

Plants have been an incredibly rich source for the identification of compounds that activate cytoprotective genes. The development of a simple microtiter-plate-based assay [19] to assess induction of the cytoprotective enzyme NAD(P)H): quinone oxidoreductase 1 (QR1) in mouse Hepa1c1c7 cells has greatly facilitated the ability to screen for and identify many cytoprotective phytochemicals. For example, a collective effort of colleagues at the University of Illinois at Chicago and Purdue University has identified 66 compounds from 18 plant species that are active in the QR1 assay [20]. Representative phytochemicals that have been shown to activate cytoprotective genes are listed in Table 7.1.

Several plant families important for human diets are particularly rich in ARE inducers. For example, many organosulfur activators have been isolated from garlic and onion, edible members of the *Allium* family. The *Cruciferae* family of vegetables, including broccoli, cabbage, Brussels sprouts, horseradish, mustard, and watercress, produces a particularly large and functionally diverse number of potent ARE inducers (Table 7.1). These plants contain glucosinolates, the thioglucoside conjugates of the ARE-activating species. Altogether, over 120 glucosinolates have been identified from various plants [21]. These are enzymatically converted to the ARE-inducing forms either by myrosinase, a thioglucosidase that is localized in a separated cellular compartment and is released upon maceration or chewing, or by intestinal microflora after ingestion [22]. Three general classes of ARE inducers produced by myrosinase-catalyzed hydrolysis of glucosinolates are isothiocyanates, indoles, and epithionitriles (Table 7.1). The enzymatic hydrolysis mechanisms involved in their release are reviewed elsewhere [23, 24]. Finally, phenolic compounds, another important class of Nrf2-activating agents, have been isolated from diverse plant families, including grapes (*Vitaceae*) and teas (*Theaseae*), which are rich sources of flavonoids and related catechins.

Importantly, *Nrf2*$^{-/-}$ mice experiments highlight the key role of Nrf2 in mediating the cytoprotective effects of the phytochemicals discussed above. Thus, sulforaphane, one of the active components of broccoli sprout extracts, has been reported to inhibit skin [12] and forestomach [10] carcinogenesis in wild-type (wt) mice, but its ability to do so is significantly attenuated in *Nrf2*$^{-/-}$ animals. In addition,

[2] Clinicaltrials.gov identifier NCT01315665.

[3] Clinicaltrials.gov identifier NCT01335971.

Table 7.1 Structures of phytochemicals that upregulate cytoprotective enyzmes through Nrf2

Structural/functional classes	Representatives	Plant sources
Chalcone, MA, phenol	Isoliquiritigenin	Licorice, shallot, tonka bean
Chalcone, MA, phenol	Xanthohumol	Hops
Cinnamate, 1,2-diphenol	Caffeic acid	Lignin-containing plants
Cinnamate, 1,2-diphenol	Ferulic acid	Apples, cabbages, plums
Diarylheptanioid, MA, 1,2-diphenol	Curcumin	Turmeric
Diarylheptanioid, MA, 1,2-diphenol	Yakuchinone B	*Alpinia oxyphylla* Miquel
Masked 1,2-diphenol	Capsaicin	Spicy peppers
Terpenoid 1,2-diphenol	Carnosol	Rosemary
Flavanol, 1,2,3-triphenol	EGCG	Green tea
1,2-Diphenol	Ellagic acid	Berries
Flavonoid mixture	Silymarin (mixture)	Milk thistle
Flavanone, masked 1,2-diphenol	Silibin (single compound)	Milk thistle
Masked 1,2-diphenol, MA	10-Shogaol	Ginger
Flavone, phenol	Apigenin	Chamomile, thyme
Flavone, phenol	Luteolin	Celery, thyme
Isoflavone, phenol	Genistein	Soy
Isoflavone, phenol	Biochanin A	Red clover
Flavonol, 1,2,6-triphenol	Quercetin	Apples, capers
Flavonol, 1,6-diphenol	Kaempferol	Apples, broccoli, tea
Flavonol, 1,2,6-triphenol	Fistein	Acacia, mangos
Flavonol, phenol	Galangin	*Alpinia officinarum*
Trans-stilbene, phenol	Resveratrol	Grape skins, berries
Trans-stilbene, 1,2-diphenol	Piceatannol	Grape skins, berries
Withanolide, MA	18-Hydroxywithanolide D	Tomatillo
Withanolide, MA	Withaphysacarpin	Tomatillo
Norwithanolide, MA	Subtrifloralactone A	*Deprea subtriflora*
Coumarin, MA (?)	Coumarin	Leguminosae spp.
Butenolide, MA (?)	β-Angelica lactone	*Archangelica officinalis*
Coumarinoid, MA (?)	Auraptene	Citrus
Terpenoid, furan	Cafestol	Green coffee beans
Terpenoid, furan	Kahweol	Green coffee beans
Terpenoid, MA	Zerumbone	Tropical ginger
Terpenoid, MA	Citral (geranial)	Lemongrass
Terpenoid, MA	Celastrol	Thunder of God vine (*T. wilfordii*)
Terpenoid (carotene)	Lycopene	Tomatoes
Phytosterol, MA	*E*-Guggulsterone	*Commiphora mukul*
Alkaloid, MA	Cryptolepinone	Sida acuta
Organosulfide	Allicin	Garlic
Organosulfide	Diallyl disulfide	Garlic
Organosulfide	Diallyl trisulfide	Garlic
Organosulfide	Dilallyl sulfide	Garlic
Organosulfide	S-allyl cysteine	Garlic
Isothiocyanate	Sulforaphane	Broccoli
Isothiocyanate	Phenethyl ITC	Turnips, watercress
Isothiocyanate	6-Methylsulfinylhexyl ITC	Wasabi
Indoles	Indole-3-carbinol > DIM[a]	Brussels sprouts, cabbages
Indoles	Brassinin	Chinese cabbage
Epithionitrile	1-Cyano-2,3-epithiopropane	Cabbages
Epithionitrile	1-Cyano-3,3-epithiobutane	Cabbages

MA Michael acceptor, *ITC* isothiocyanate

[a] As described in the text, indole-3-carbinol is a precursor for diindolylmethane (DIM), which is considerably less toxic

sulforaphane was able to protect the blood–brain barrier post injury only in wt Nrf2 mice [13]. Similarly, resveratrol, a flavonoid-like molecule produced by many plants, protected against high-fat-diet-induced oxidative stress in aortas of wt but not *Nrf2*$^{-/-}$ mice [9].

7.3.2 The Chemistry Required for Phytochemicals to Activate Nrf2

7.3.2.1 Reactivity with Thiolates

Despite the high level of overall structural diversity among ARE inducers, many cytoprotective phytochemicals from different plant sources share common thiol-reactive chemical motifs (collectively shown in Fig. 7.1 as red moieties). For example, the presence of an α,β-unsaturated carbonyl group, a potential Michael acceptor, is a particularly common feature in ARE inducers, including withanolides (e.g., withaphysacarpin), chalcones (e.g., isoliquritigenin), butenolides (e.g., β-angelica lactone), oxidized terpenoids (e.g., zerumbone, *E*-guggulsterone, and citrals), and curcuminoids (e.g., curcumin). Other inducer classes are similarly electrophilic epithionitriles (e.g., 1-cyano-2,3-epithiopropane), isothiocyanates (e.g., sulforaphane), and organopolysulfides (e.g., allicin). Talalay and colleagues in 1988 first recognized that the diverse structures share the ability to react with thiolate groups [25]. They hypothesized that gene induction takes place by virtue of an intracellular sensor that contains one or more reactive cysteine residues, modification of which by inducing agents would lead to target gene activation. This seminal hypothesis is supported by numerous studies that followed, which directly linked the presence of particular functional groups to Nrf2–ARE induction, as shown with zerumbone [26], chalcones [27], flavonoids [28], and withanolides [20, 29]. Furthermore, a strong correlation has been identified between inducer potencies and chemical reactivities toward thiolates [30].

Finally, in 1999 a key repressor of Nrf2 was discovered, the Keap1 protein, which was found to possess an unusually large number of cysteines (25 and 27 in the mouse and human proteins, respectively) [31]. As described in detail in Sect. 7.3.4.1, a subset of these cysteines, C151 in particular, has been found to be important for Nrf2 activation by phytochemicals.

7.3.2.2 Phytochemicals that Do Not Have the Ability to React with Thiolates

While the critical feature of an Nrf2 activator appears to be the ability to react with thiolates, a large number of phytochemicals, including phenols, monosulfides, furans, and indoles, would need to acquire this property through metabolic and/or chemical processing. Studies on the biotransformations of many such molecules are surprisingly limited, despite the current interest in the health benefits of phytochemicals. Herein, we summarize the available experimental evidence from the literature implicating mechanisms of converting phytochemicals to thiol-reactive species. We also discuss two classes for which transformation to a thiolate-reactive species is considerably more difficult to ascertain, carotenoids and 1,3-polyphenols.

Electron-rich phenolic compounds featuring "additive" distribution of electron-donating groups are one relatively well-characterized chemical class of phytochemicals that require electrophilic conversion for inducer activity (Fig. 7.1, blue structural fragments). These are compounds that contain phenolic hydroxyl groups in a conjugated system with an even number of carbons separating them. Representative examples of this class include 1,2-diphenols (e.g., epigallocatechin gallate (EGCG), caffeic acid, and piceatannol) and vinylogous 1,6-diphenols, with an alkene spacer acting as the "electron conduit" (e.g., quercetin and kaempferol). The 1,2- 1,4-, and 1,6-diphenols, but not the mono- or 1,3-dihydroxy variants, can be readily oxidized in vivo by a variety of mechanisms (Fig. 7.2) to the corresponding quinoids and will be referred to herein as quinoid-forming phenols. In fact, such phenols have been associated with induction of carcinogen-detoxifying enzymes by one of the earliest observations in the field [32]. The resulting Michael acceptors can then readily react with cysteine thiolates (Fig. 7.2). For example, the well-characterized flavonoid quercetin is expected to be a very weak electrophile intrinsically due to the electron-donating effect of the 3-hydroxyl group and both aromatic rings (see Fig. 7.3a for numbering scheme). However, in human blood plasma, quercetin is readily oxidized to a significantly more electrophilic quinone methide [33] (Fig. 7.3a). Importantly, the oxidized species is much more reactive toward thiolates, and its conjugation products have been detected with GSH, N-acetylcysteine (NAC), cysteine [34, 35], and protein cysteine residues [36]. Similarly, EGCG contains several aromatic 1,2-dihydroxy units, and thus can form quinones that react with isolated and protein thiolates, as demonstrated in both biochemical experiments and cells [37, 38]. EGCG has also been found in mouse urine as the S-cysteinyl–EGCG conjugate after a high oral dose [39]. Furthermore, the oxidative conversion of phenolic compounds to Michael acceptors has been shown to correlate strongly with cytoprotective enzyme induction in studies evaluating *tert*-butylhydroquinone (tBHQ) [40] and EGCG analogs [38], as well as a broad series of phenols [41]. As further evidence that the oxidation of phenols is a prerequisite to ARE induction, Cu^{2+}or other oxidized transition metal cations in the media strongly stimulated the ARE induction potential of *para*- and *ortho*-hydroquinones [42]. These metal ions act as catalysts in the oxidation of phenols to Michael reaction acceptors (Fig. 7.2) under aerobic conditions. Importantly, transition metal salts had no effect on inducer activity of the corresponding quinones or sulforaphane [43].

The organosulfur compounds from garlic and onions were shown by the Wattenberg group in 1988 to have interesting structural requirements for inducer activity [44]. Garlic organopolysulfides and derivative thiosulfinates, such as allicin (Fig. 7.1), were able to induce the cytoprotective enzyme glutathione S-transferase (GST), as might be expected from their ability to modify cellular thiols. However, the monosulfide diallylsulfide (DAS) is also an inducer, and the mechanism by which it activates Nrf2 is yet to be delineated. In addition, the diallyl forms of the

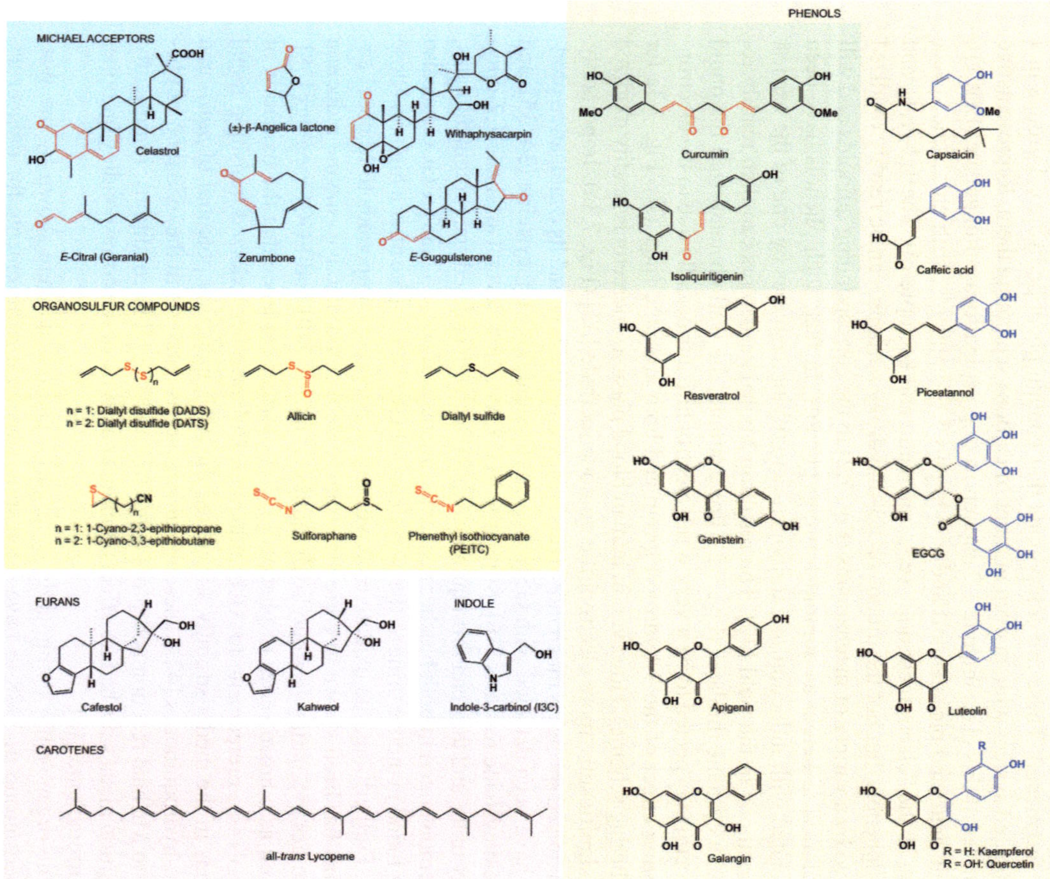

Fig. 7.1 Structures of phytochemicals arranged by the chemistry that leads to Nrf2 activation. All compounds shown activate Nrf2. Thiol-reactive chemical motifs are shown in *red*, and "additive" distribution of electron-donating groups, which can be oxidized to quinoids, are shown in *blue*

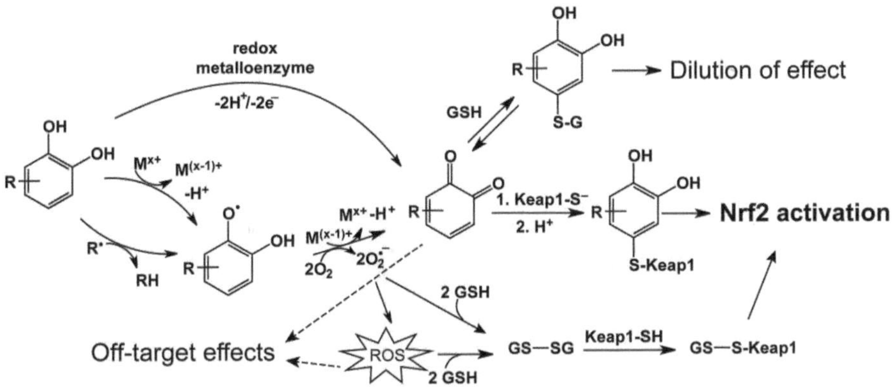

Note: Keap1 thiols in scheme could be replaced with those of other sensor proteins: PTEN C124, etc.

$R^{\bullet} \rightarrow RH$ =*DIRECT ANTIOXIDANT EFFECT (Free radical trapping)*

$O_2 \rightarrow O_2^{\bullet-}$ =*PROOXIDANT EFFECT (ROS generation)*

Nrf2 Activation =*INDIRECT ANTIOXIDANT EFFECT (Upregulation of cytoprotective genes)*

Fig. 7.2 The three roles a quinoid-forming polyphenol (represented by the R-catechol) can play: prooxidant, direct antioxidant, and indirect antioxidant. As a *direct antioxidant*, in the presence of a high concentration of free radical species, a polyphenol can trap the radical, forming a relatively stable radical species. As a *prooxidant*, in the presence of catalytic amounts of a transition metal, a polyphenol can promote the formation of superoxide and other ROS, en route to formation of a Michael acceptor. An alternate path to oxidation of the polyphenol is catalyzed by a metalloenzyme and occurs without the production of ROS. Once the quinoid group is formed, the Michael acceptor group can react with a thiolate molecule. There is evidence that a quinone reacts with a key Keap1 sensor cysteine, C151, leading to Nrf2 activation, as described in the text. Upon activation, Nrf2 upregulates a battery of antioxidant enzymes and other cytoprotective enzymes, known as the *indirect antioxidant* effect. Reaction of the quinoid with GSH and subsequent elimination from the cell will lead to dilution of the effect. Alternative mechanisms of Nrf2 activation by radicals or ROS not depicted are oxidation of sensor cysteines, or formation of disulfides among sensor cysteines

di- and tri-sulfides were much more potent than the propyl version. In surmising why the diallyl sulfides might have higher potency, it is interesting to note that cytochrome P450 2E1 (CYP2E1) is implicated in sequential conversion of the sulfide to the corresponding sulfoxide (DASO) and sulfone ($DASO_2$) derivatives [45]. Further oxidation of the sulfone metabolite generated an electrophile that was shown to act as a suicide inhibitor of CYP2E1, as well as other unidentified cytochromes implicated in bioactivation of various cytotoxins [45]. While the exact mechanism of this inhibition remains to be established, the irreversible nature of the antagonism implies generation of a reactive intermediate capable of covalent modification of the enzyme involving, in all likelihood, an active-site cysteine [46]. In addition, liquid chromatography–mass spectrometry (LC–MS/MS) analysis of bile fluids from rats treated with DAS, DASO, and $DASO_2$ identified several GSH conjugates, implicating epoxidation of the allylic group as an important metabolic activation step for all three compounds [47]. Unlike the epoxides of DAS and DASO, the $DASO_2$-

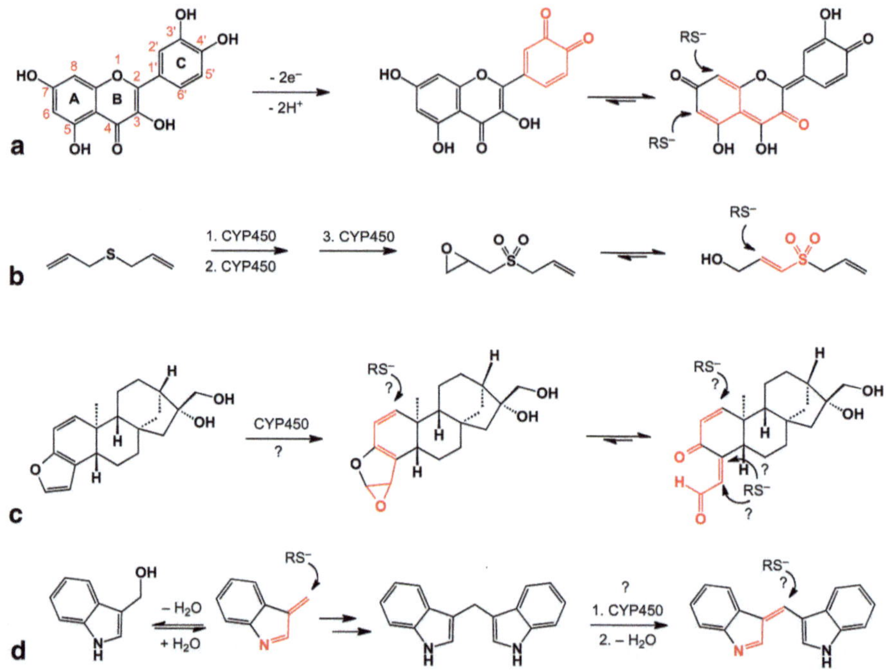

Fig. 7.3 Biotransformations implicated in conversions of quercetin (**a**), DAS (**b**), coffee triterpenoids (kahweol shown here) (**c**), and I3C and DIM (**d**) into thiol-reactive conjugated electrophiles: quinoids (quercetin), sulfone (DAS), epoxide or γ-ketoenal (kahweol), and indolenines (I3C and DIM). The established or proposed reactive groups are highlighted in *red*

derived epoxide provides a unique entry into a sulfone-activated Michael acceptor containing a sulfonylprop-2-en-1-ol group (see Fig. 7.3b for the proposed mechanism). This so-called "soft" electrophile, with highly distributed charge density, is much more likely to react with a thiolate, a similarly polarizable "soft" nucleophile, rather than the "hard" epoxide (see Sect. 7.3.3 for a brief discussion of hard and soft reagents and other factors affecting electrophile chemoselectivity). Therefore, oxidation of the sulfide, in combination with epoxidation of the allylic group, could be responsible for the inactivation of CYP2E1 and, possibly, induction of Nrf2 by modifying sensor cysteines.

Furan-containing compounds, such as the diterpenes cafestol and kahweol, are known to act as the principal Nrf2 activators in coffee despite lacking a thiol-reactive functional group [48, 49]. Although the exact mechanism of induction is yet to be established, a recent mass spectrometric study of bile fluids from mice injected with cafestol was interpreted by the authors to suggest that epoxidation is the key step in converting coffee furans into thiol-reactive species [50]. However, the study did not distinguish between direct addition to the epoxide and addition at a remote double bond conjugated to the epoxide or Michael addition to a γ-ketoenal, arising from a ring-opening reaction of the oxidized furan (depicted in Fig. 7.3c).

The distinction is important because furan epoxides and corresponding dicarbonyl derivatives have long been associated with severe cytotoxicity of furan-containing compounds [51], due to both high reactivity and a tendency to react with oxygen and nitrogen nucleophiles, in addition to thiolates. This, however, is inconsistent with the generally cytoprotective properties of the coffee furans. The particular structural environment of electrophiles derived from these furans may account for the observed shift in the balance of toxic and protective effects (Fig. 7.3c). Thus, the kahweol epoxide presents a doubly conjugated epoxide (epoxydiene) requiring a soft thiolate nucleophile to attack the soft electrophilic center next to a highly sterically congested quaternary carbon. The corresponding ketoenal is also deactivated toward nonspecific additions by both cross-conjugation of the dienone functionality and β,β-dialkyl substitution of the unsaturated aldehyde [52]. Alternatively, participation of less reactive metabolites, such as conjugated lactones, for example, which have been detected as metabolic derivatives of furan-containing compounds [53], may account for the low toxicity of these phytochemicals. Further studies will be required for understanding the unique properties of these furans.

Indole-3-carbinol (I3C), a glucosinolate breakdown product, and its digestive product 3,3′-diindolylmethane (DIM) are ARE inducers [54] and have many other cancer chemopreventive effects [55]. However, while DIM has been described as an effective inducer of the ARE with little associated long-term toxicity [56], I3C has a weak level of Nrf2 activation and is associated with a complex interplay of pro- and anticarcinogenic effects (for a review see [56]). Although no specific mechanism accounting for the ARE induction ability of DIM has been established, reaction schemes for producing electrophilic species capable of covalent modification of thiols could be inferred from known metabolic and chemical transformations of I3C, DIM, and 3-methylindole (3MI), a by-product of tryptophan metabolism by intestinal microflora [57]. In the case of the last compound, a highly electrophilic thiolate-reactive derivative, 3-methyleneindolenine, an α,β-unsaturated imine, is produced via cytochrome P450 (CYP)-mediated dehydrogenation [58]. This species is postulated to be responsible for the cytotoxicity of this substituted indole [58, 59]. The reactivity of the α,β-unsaturated imine toward nucleophiles is promoted by a concomitant rearomatization of the indole nucleus, which makes it a rather nondiscriminating and thus toxic agent, capable of modifying a variety of functionalities found in a cellular environment. 3-Methyleneindolenine also serves as an intermediate in the dehydration–conjugate addition–retro-aldol cascade in the acid-catalyzed conversion of I3C to DIM [60] (Fig. 7.3d). Therefore, the low level of Nrf2 activation and pro- and anticarcinogenic effects associated with I3C could be explained by the formation of both the toxic electrophile and DIM, respectively, in the course of the same chemical process. On the basis of established biotransformations of DIM [61], it is tempting to propose that an oxidation event similar to that seen for 3MI can lead to the extended conjugation-stabilized Michael acceptor, 3-(3-indolylmethylene)-indolenine (Fig. 7.3d). We must note that no thiol conjugates of DIM metabolites have been isolated, perhaps, due to reversibility of such additions to the conjugation-stabilized electrophile (see Sect. 7.3.3 for further discussion of the importance of reversibility). Significantly, a sulfate-conjugated hydration product of the proposed species has been isolated as one of the major biotransformants from cultured cancer cells [61]. The Michael acceptor produced by both I3C and 3MI may be the principal agent responsible for associated toxicities. Unlike 3MI, however, the biotransformation of I3C can also lead to DIM, a more stable indole derivative [62] that is unlikely to produce a nondiscriminating electrophile. This example underscores the fact that the level of reactivity of phytochemically derived electrophiles could be the key determinant in a sensitive balance of cytotoxic and cytoprotective effects.

Carotenoids, lycopene in particular, have been shown to activate the ARE via Nrf2 [63]. The carotenes are unsaturated hydrocarbons and thus contain no thiol-reactive species. However, the authors point out that the ethanolic extract of the lycopene preparation, containing unidentified hydrophilic lycopene derivatives, activated the ARE with a similar potency as lycopene. Therefore, it seems likely that oxidation of the polyene, leading to formation of an electrophilic species (e.g., citral [64]), is a prerequisite for Nrf2 activation.

In addition, there are a handful of phenols (non-quinoid forming) that have no electrophilic moieties, or facile nonenzymatic autoxidative paths to obtaining these moieties, which have been shown to activate cytoprotective enzymes, such as resveratrol [65] and galangin [28]. The Nrf2-dependence of ARE activation by resveratrol has been particularly well established in studies in *Nrf2*$^{-/-}$ mice [9] and normal human small airway epithelial cells (SAECs) [66]. Resveratrol and other related phenols can be hydroxylated to quinoid-forming species by a member of the cytochrome P450 (CYP) superfamily of monooxygenases, such as pro-carcinogen activating CYP1A1, CYP1A2, and CYP1B1 [67, 68]. These enzymes are transcriptionally controlled by the aryl hydrocarbon receptor (AhR). So-called bifunctional ARE inducers, such as β-naphthoflavone (β-NF), activate Nrf2 by first binding to and activating AhR, which in turn leads to upregulation of CYP enzymes [69]. However, resveratrol is a known inhibitor of AhR [70, 71], and therefore in uninduced normal cells resveratrol may not be hydroxylated by CYPs prior to Nrf2 activation. Resveratrol activation of QR1 through the estrogen receptor β (ERβ) in breast cancer cells has been explored as another mechanistic possibility. Various studies support a model in which binding of phytoestrogens, including resveratrol, to ERβ causes ERβ to bind and activate the QR1 ARE [72–75]. However, this mechanism appears to be restricted to cancer cells overexpressing ERβ [75]. A third mechanism has been suggested based on the observation that resveratrol binds to and inhibits quinone reductase 2 (QR2) with low nanomolar affinity [76]. QR2's function is not clearly elucidated, but it is known to catalyze the reduction of quinones, among several other classes of electron-deficient compounds. The authors suggested that resveratrol may activate Nrf2 by inhibiting QR2, resulting in the accumulation of endogenous quinones that can then induce electrophilic stress by modifying cellular thiols [76]. A similar indirect induction mechanism could apply to other phytochemical inhibitors of QR2, including quercetin, kampferol, apigenin, or genistein, that inhibit this reductase at physiologically relevant concentrations [76, 77]. This hypothesis warrants further investigation. Finally, a fourth mechanism accounting for activation of Nrf2 by non-quinoid-forming phenols by means of kinase activation is explored for genistein in Sect. 7.3.4.3.

7.3.2.3 Role of ROS in Nrf2 Activation by Phenolic Compounds

One important area of consideration is to what extent ROS are involved in the activation of Nrf2 by phytochemical inducers. Phenolic ARE inducers can play multiple roles in the redox status of a cell (Fig. 7.2). They are able to act as both direct antioxidants, scavenging free radicals, and, if converted to thiolate-reactive electrophiles, as indirect antioxidants by upregulating antioxidant genes [78]. Depending on the experimental conditions, they can also act as pro-oxidants [79]. As shown in Fig. 7.2, the nonenzymatic autoxidation of a polyphenol can lead to ROS formation. This generation of ROS may play a role in Nrf2 activation, as well as have potential off-target effects such as toxicity. The ability of ROS to upregulate Nrf2 is well established, for example by treatment of cells with H_2O_2 [80]. One mechanism by which ROS may activate Nrf2 is depicted in Fig. 7.2. In this scenario, the generation of superoxide and other ROS can lead to oxidation of GSH to GSSG [81]. GSSG then could modify sensor thiolates such as Keap1 cysteines [82, 83], thereby activating Nrf2 (see also Sect. 7.3.4.1). In addition, sensor thiolates can be directly oxidized by ROS (reviewed in [84] and [85]).

There are indications that ROS mediate signaling for some ARE inducers. For example, the generation of ROS in cells [86, 87] and cell media [88] by EGCG has been well established. Importantly, ROS scavengers NAC, GSH, superoxide dismutase (SOD), and catalase all inhibited the induction of heme oxygenase 1 (HO-1) by EGCG, as shown in bovine aortic endothelial cells (BAECs) [89]. Thus, an EGCG-induced ROS, rather than an EGCG-derived electrophile, mediated EGCG-induced HO-1 expression under these conditions. A role for ROS as a secondary messenger in Nrf2 activation, specifically H_2O_2, has also been shown for a different ARE inducer class, dithiolethiones, in Hepa 1c1c7 cells [80]. In addition, a role for ROS in Nrf2 activation by 21 flavonoids (including fisetin, kaempferol, and quercetin) was explored [90]. A high level of correlation was observed between the flavonoids' ability to activate the ARE and their computed energy levels of the highest occupied molecular orbitals (E_{HOMO}), representing the tendency of the flavonoid to donate electrons in redox processes, for example. Thus, more oxidizable flavonoids possessing less negative E_{HOMO} values were generally more potent inducers of ARE-mediated gene expression. Therefore, ROS may be important secondary messengers for flavonoids.

Interestingly, we note that there are three outliers in the flavonoid correlation analysis [90] with much greater abilities to activate the ARE than predicted by their E_{HOMO} values (and hence their tendency to donate electrons in the formation of ROS). This boost in induction potency could be associated with a particular set of structural features. Thus, only these three flavonoids (quercetin, morin, and myricetin) out of 21 tested can be predicted to form extended conjugation-stabilized quinone methides involving the C ring, as products of a two-electron oxidation sequence (Fig. 7.3a). We hypothesize that while generation of ROS by flavonoids may contribute to activation of Nrf2, the ability to form a conjugation-stabilized electrophile, which can react with thiolates such as quinoid, produces a much greater extent of ARE activation due to its ability to react directly with sensor thiolates, such as Keap1 cysteines. Importantly, the quinone methide derived from quercetin has a remarkably high stabilization effect of the extended conjugation, as shown by its ability to form reversible thiol adducts [34]. Finally, in support of this hypothesis, out of a series of five structurally similar flavonoids, only the quinoid-forming ones (kaempferol, quercetin, and luteolin) were able to react with a thiol (GSH) to form *mono*- and *bis*-GSH conjugates, without forming radicals and ROS [91, 92]. In contrast, the other two flavonoids (apigenin and naringenin), which are not able to form conjugation-stabilized electrophiles, generated radicals and ROS and could not form conjugates with GSH.

In considering whether generation of ROS is a secondary messenger in Nrf2 activation by flavonoids and other phytochemicals, it is important to note that the two-electron oxidation of phenols need not necessarily lead to ROS formation. If a polyphenol can be recognized as a substrate of an oxidizing metalloenzyme, it can be converted to the Michael acceptor without production of ROS (Fig. 7.2). To our knowledge, this theory remains to be tested. Dosage is likely critical as to whether useful or harmful levels of ROS are produced by treatment with ARE inducers [93]. Much work remains to determine the role(s) that ROS play in Nrf2 activation by phytochemicals, including whether the ROS generated by therapeutic and physiologically relevant concentrations of phytochemicals have deleterious off-target effects, or are relatively harmless and perhaps relevant for participating in numerous signaling mechanisms, which are beyond the scope of this Perspective (reviewed in [85, 94, 95]). The dosage amount is likely very important, as illustrated through the example of the synthetic oleanane triterpenoid CDDO-Im. While at low concentrations ($\leq$100 nM) CDDO-Im is a potent Nrf2 inducer with undetectable adverse effects, above 300 nM ROS-mediated toxicity is observed [96]).

7.3.3 Basis of Phytochemicals as Therapeutic, Rather than Toxic, Agents

The cytoprotective roles of phytochemicals discussed thus far could be considered surprising, given that most if not all were produced for biodefense against insects, bacterial parasites, and other animals, including humans, consuming plant parts. The question arises then as to how these molecules have a cumulatively cytoprotective effect. One critical factor is dosage, as at high concentrations most of these molecules have been shown to display at least some level of toxicity. At low concentrations then, the residual "toxicity" maintains cells in a state of adaptive stress, providing them with tools for counteracting a variety of adverse conditions [97]. The ability of a phytochemical to induce such a state selectively plays a critical role in the cytotoxicity/cytoprotection balance. This balance appears to be rather sensitive to subtle changes in chemical structure. Two considerations that appear to be significant are selectivity for stress-sensing cysteines and reversibility of thiol modification.

There are various factors that contribute to the preference of an electrophile for stress-sensing cysteines. First, cytoprotective compounds (or their bioactivated derivatives) contain "soft" electrophilic centers, which generally favor an attack by a corresponding soft nucleophile, best represented in a cellular environment by a thiolate anion. In phytochemicals, soft electrophilicity is associated with a carbon center conjugated through a network of π-bonds to a reactive functional group, such as a carbonyl (Fig. 7.3a), sulfone (Fig. 7.3b), epoxide (Fig. 7.3c), or imine (Fig. 7.3d), or with an electrophilic center activated directly by large, polarizable atoms (e.g., epithionitriles, isothiocyanates, etc.). The molecular orbital effects at the root of these preferences are important but are beyond the scope of this Perspective (see [98] for relevant discussion). Soft electrophilicity, however, is not sufficient for shifting the therapeutic window toward cytoprotection, as highly reactive albeit soft agents will modify unintended thiols (i.e., hemoglobin, human serum albumin, GSH, etc.) [3] and even non-thiol targets (i.e., nucleic acids).

Second, stress-sensing cysteines are likely maintained by their immediate environment in a highly reactive state. This could include pK_a reduction [99] and the presence of Brønsted acids to orient and activate the incoming electrophiles through hydrogen bonding [52] (see Fig. 7.4a). It is highly likely that one or several Keap1 cysteines are in fact presented in such an environment (see Sect. 7.3.4.1).

Third, the structural context of presenting the electrophilic functionalities is likely to play an essential role in an electrophile's ability to react selectively. Thus, another contributing factor noted in enhancing reactivity with sulfur nucleophiles is the presence of a neighboring group, such as a hydroxyl functionality, in the vicinity of a reactive β-carbon of cinnamates, chalcones, curcuminoids, bis(2-hydroxybenzylidene) derivatives [30], and some withanolides [20]. A neighboring hydroxyl can guide an incoming thiolate anion via a transient charge–dipole interaction (hydrogen bond) resulting in a highly selective process that enhances the effectiveness of such agents (Fig. 7.4b).

In addition, a sterically hindered electrophile may exercise a superior degree of selectivity toward uniquely reactive thiolates (e.g., Keap1 cysteines). For example, both the high effectiveness and low toxicity profile of synthetic oleanane triterpenoids have been ascribed to a special combination of the high intrinsic electrophilicity of the Michael acceptor activated by both keto and cyano groups (Fig. 7.4c) and sterics, specifically the presence of a Michael acceptor functionality directly adjacent to a highly congested quaternary carbon center [100] (Fig. 7.4d). Congestion in the vicinity of the reactive β-carbon reduces the reactivity of the Michael acceptor by crowding the transition state, where trigonal planar geometry is converted into a more sterically demanding tetrahedral state. This should allow only the most reactive thiolates to overcome the steric barrier and form covalent adducts. It must be noted, however, that mere reduction of reactivity without adjustment of other factors is expected to result in a situation where the enhanced selectivity and reduced toxicity come at the expense of potency. Thus, of two stereoisomers of guggulsterone, a terpenoid that lacks any known toxicity, only the *E*-form displays moderate ARE induction [101], due to, presumably, a poor accessibility by nucleophiles of the trisubstituted α,β-unsaturated ketone furnished with a quater-

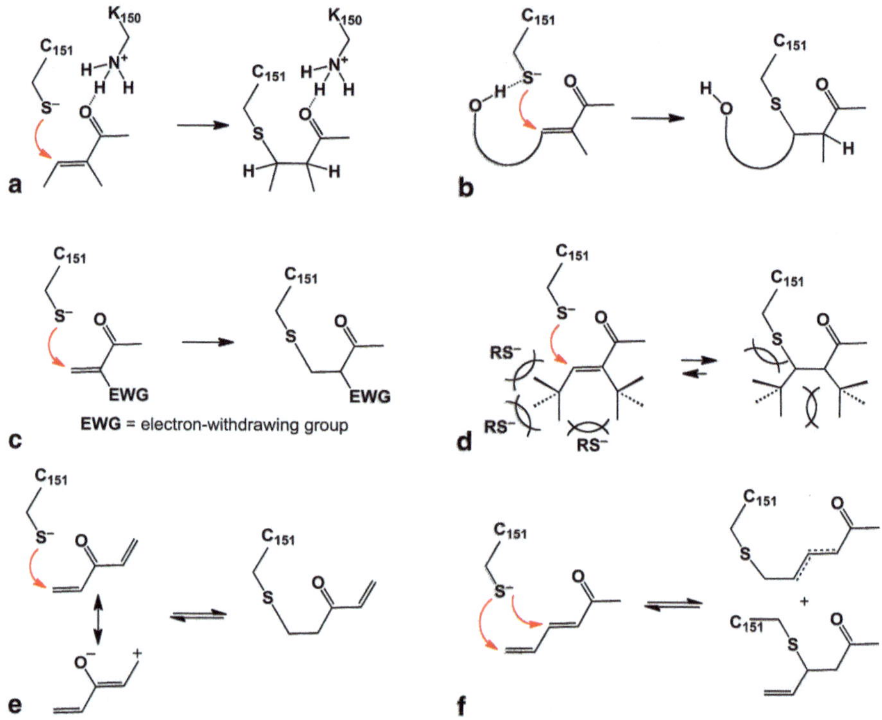

Fig. 7.4 Structural features of Michael acceptors affecting the rates of forward and reverse reactions with thiols. Keap1 C151 is shown as an example thiol. **a** Brønsted acid catalysis by a neighboring residue (Keap1 K150). **b** Neighboring group (proximity) catalysis through hydrogen bonding with a hydroxyl adjacent to the β-carbon. **c** Alkene polyactivation, favoring addition. **d** Steric congestion, preventing additions by less reactive nucleophiles via transition state crowding. **e** Cross-conjugation and **f** extended conjugation stabilize an electrophile by diluting the partial positive charges at the electrophilic centers and thereby reducing its reactivity and promoting reversibility

nary center as the α-carbon substituent. The relatively low toxicity of coffee furans, as compared to other furans, can also be explained by steric factors, if the operative electrophile is a conjugated epoxydiene formed directly from kahweol (Fig. 7.3c) or after oxidative processing of cafestol.

The reversibility of addition, the fourth factor on our list, is emerging as a crucial characteristic of highly effective inducers with low associated toxicity [102]. Clearly, the entropically favored reversal of electrophilic additions is expected to (1) alleviate many potentially adverse outcomes associated with unintended alkylation events, (2) activate multiple stress sensors at a greatly reduced concentration, and (3) exhibit catalytic turnover (recycling) in maintaining the state of adaptive stress. ARE inducers must compete with prominent blood proteins—hemoglobin and serum albumin—and GSH, among many other potential nucleophiles. GSH, the most available cellular thiol responsible for detoxification, can be conjugated with electrophiles in an uncatalyzed manner, or the conjugation can be catalyzed by

GSTs (reviewed in [103]). Once formed, if sufficiently stable, these conjugates are excreted from the cell by transmembrane multidrug resistance-associated proteins (MRPs), leading to dilution of the ARE inducer and the signaling event (Fig. 7.2). Thus, highly reversible ARE inducers may escape elimination from the cell by deconjugating from GSH prior to being transported by an MRP, maintaining their cellular concentration more effectively than ARE inducers that form much more stable conjugates. For "reversible" ARE inducers, conjugation with GST has been hypothesized to be an intermediate step [104], leading via continuous deconjugations to activation of less represented but highly reactive "sensory" thiolates, responsible for antioxidant effects (Fig. 7.2).

For example, the reversible addition of sulforaphane to thiols has long been suspected to be responsible for the pleiotropic nature of its cellular activity [78, 105]. In addition, the dramatically enhanced Nrf2-induction potential of CDDO and its variants over traditional phytochemical inducers may be due in large part to the facile reversibility of their thiol conjugates [100]. While a systematic analysis of chemical features responsible for the reversibility of thiol conjugations would benefit the field, certain generalizations about these features can be made herein. Thus, unusual stabilization of the electrophile through cross-conjugation (Fig. 7.4e) or extension of conjugation (Fig. 7.4f) could be invoked in a few cases documenting the favorability of addition reversal [34, 52, 100]. Both effects should lead to the enhanced delocalization of the partial positive charge induced by the associated carbonyl, reducing the electrophilicity of β- (δ-, etc.) carbon centers. For example, in celastrol, a quinone methide triterpene that has submicromolar potency for ARE activation, extended conjugation can account for the observed reversibility of its reaction with GSH [106]. Similarly, reversibility due to enhanced stabilization of the electrophile could explain the reduced toxicity and unique effectiveness of quercetin, when compared to structurally related flavonoids and flavonols in particular. Thus, the glutathionyl adducts of the quercetin-derived quinone methide have been shown to be reversible on the order of minutes to hours [34]. In addition, we suggest that the reversibility of addition may play an important role in shifting the effectiveness-to-toxicity balance in yet-to-be established cases. For example, the divergent toxicity profiles of I3C and DIM (see Sect. 7.3.2.2) could be explained by the reversibility of the DIM-derived thiol conjugate due to conjugation extension by a second aromatic group.

Alternatively, the destabilization of an adduct through steric congestion at positions adjacent to the nascent C–S bond may also facilitate elimination, as carbon hybridization changes from trigonal planar to tetrahedral. Thus, we hypothesize that the unusually low toxicity of CDDO [100] is likely the result of addition reversibility due to steric congestion in the corresponding adduct (Fig. 7.4d). In contrast, ARE inducers that do not form relatively reversible adducts, such as the EGCG *o*-quinone, for example, may be eliminated from the cell rather quickly through sequestration by GSH and subsequent elimination by an MRP. For these compounds, generation of superoxide and ROS as signaling intermediates may represent a more effective pathway for Nrf2 activation (Fig. 7.2).

7.3.4 Mechanisms of Nrf2 Activation

7.3.4.1 Modification of Keap1 Cysteines: Identification of Key Sensors for ARE Inducers

A key mechanism by which phytochemicals activate Nrf2 is revealed by their common ability to react with cysteine residues. As mentioned in Sect. 7.3.2.1, the Nrf2-repressor protein Keap1 is particularly cysteine-rich. Keap1 directly binds to Nrf2 [31] and represses the transcription factor in at least two ways. First, Keap1 contains a Crm1-dependent nuclear export signal sequence that appears to be important for maintaining Nrf2 primarily in the cytoplasm under basal conditions [107–109]. In addition, and likely most important for phytochemical signaling, Keap1 serves as a bridge between Nrf2 and the Cul3-E3-ubiquitin ligase complex, leading to the ubiquitination of Nrf2 and subsequent degradation by the 26S proteasome (reviewed in [110]). There are two separate sites in Nrf2 that bind to Keap1, termed the ETGE and DLG motifs [111]. Keap1 forms a homodimer through its N-terminal BTB domain, and it is the C-terminal Kelch domain of Keap1 that binds to the Nrf2 ETGE and DLG motifs. The seven lysines of Nrf2 that are targets for Cul3-mediated ubiquitination are located between the ETGE and DLG motifs [112]. Binding of the two Kelch domains of a homodimer of Keap1 to the ETGE and DLG sites of an Nrf2 protein molecule is believed to be critical for presentation of those lysines to an E2 ligase for ubiquitination [113]. Ubiquitination-directed degradation maintains the Nrf2 protein at a low level under normal conditions. Upon exposure to stress or compounds that modify cysteines, Nrf2 escapes Keap1 repression, and thereby is no longer ubiquitinated and degraded. This in turn leads to Nrf2 accumulation, resulting in activation of ARE-regulated genes.

Two primary methods have been used to illustrate which of the Keap1 cysteines (27 in human Keap1) are implicated in sensing ARE inducers: overexpression of Keap1 containing cysteine mutated to serine or alanine in mammalian cell lines or zebrafish, and mapping of cysteine modification in the purified protein by peptidic digestion and mass spectrometry. These complementary methods have revealed that a "cysteine code" may exist, in which particular types of ARE inducers react most readily with particular Keap1 cysteines. Cysteines 226 and 613 were shown to sense heavy metals [114], while C273 and C288 have been implicated in sensing alkenals, some cyclopentenone prostaglandins, and nitro-fatty acids [114–116]. Keap1 C151 has been shown to be required for sensing many electrophilic ARE inducers [115, 117–121]. For example, a potent imidazole derivative of CDDO (CDDO-Im), tBHQ, a quinone-forming phenol, and sulforaphane are all highly dependent on C151 to upregulate the ARE [107, 115, 120]. Finally, the importance of Keap1 C151 in sensing ARE inducers is illustrated by the Keap1 C151S transgenic mouse [122]. This mouse is not only less responsive to tBHQ for upregulation of cytoprotective enzymes but the basal levels of these enzymes are also repressed [122]. Thus, Keap1 C151 emerges as the key sensor for both endogenous agents as well as phytochemicals. The role of this cysteine as a sensor has been proposed to have evolved as a means of sensing nitric oxide [114]. Interestingly, mutation of C226 and C623 to serine has no effect on ARE activation by sulforaphane or tBHQ [114]. Therefore, C151 is currently the only known Keap1 sensor cysteine for a phytochemical, sulforaphane, and a synthetic quinoid-forming polyphenol, tBHQ.

The role of Keap1 C151 as a principal sensor for phytochemicals is further supported by peptide-mapping studies. Keap1 C151 is the only cysteine consistently and highly modified by all phytochemicals tested thus far: isoliquiritigenin, 10-shogaol, xanthohumol [123], and sulforaphane [124] (Fig. 7.5). There have been discrepant results for modification of Keap1 cysteines, in particular C151, by sulforaphane. Initially, peptide-mapping studies of human Keap1 cysteines modified by sulforaphane found C151 to be one of the least readily modified cysteines (Fig. 7.5) [125]. Based on the high dependence of sulforaphane ARE activation on C151, as well as the known labile, reversible nature of dithiocarbamates [126], i.e., the products of sulforaphane reaction with thiols, further experiments were conducted using a streamlined method to limit the reverse reaction after labeling Keap1 cysteines with sulforaphane [124]. Under these conditions, C151 emerged as one of the four most readily modified cysteines of Keap1. The adduction of Keap1 C151 by sulforaphane was also observed indirectly for Keap1 overexpressed in Cos1 cells using a biotin-switch technique, where C151 appeared to be modified to a greater extent than other cysteines [114]. The entire pattern of modification of Keap1 by sulforaphane generally shows large variability depending on the method used (Fig. 7.5 and [124]), in particular as to whether C151 is detected as modified, illustrating the importance of considering the reversible nature of electrophile–cysteine adducts. Interestingly, several cysteines including C77, C226, C368, and C489 (Fig. 7.5) were detected as readily modified regardless of the method used, indicating that certain dithiocarbamate–cysteine adducts are much more stable than others. The C151–dithiocarbamate adduct, on the other hand, is both one of the most reactive and reversible. Modeling of the BTB domain of Keap1 containing C151 [114, 120] depicts various residues nearby, including K131, R135, K150, and H154, that could participate both in lowering the pK_a of C151 and in providing acid/base catalysis for promoting both the forward (addition) and the reverse (elimination) reactions. As a confirmation, a single mutation of K150 to threonine significantly reduces the response to C151-dependent ARE inducers [115].

While the modification pattern of quinoid-forming polyphenol phytochemicals has not yet been assessed, the oxidized form of tBHQ was found to react with Keap1 C151, shown directly for purified Keap1 [127] (Fig. 7.5) and indirectly for Keap1 overexpressed in Cos1 cells [114]. Interestingly, C151 is not found to be readily modified by the oxidized form of GSH, GSSG [82] (Fig. 7.5). Phytochemicals for which ROS production is a key means of activating Nrf2 (Sect. 7.3.2.3) likely activate Nrf2 by first generating GSSG (Fig. 7.2). Thus, their Nrf2 activation would presumably be much less dependent on Keap1 C151.

While C151 is the cysteine most readily modified by phytochemicals based on current knowledge, other cysteines are sites of phytochemical addition in purified Keap1 protein (Fig. 7.5). For example, C319 is readily modified by all four phytochemicals shown in Fig. 7.5 as well as oxidized GSH. To our knowledge, this

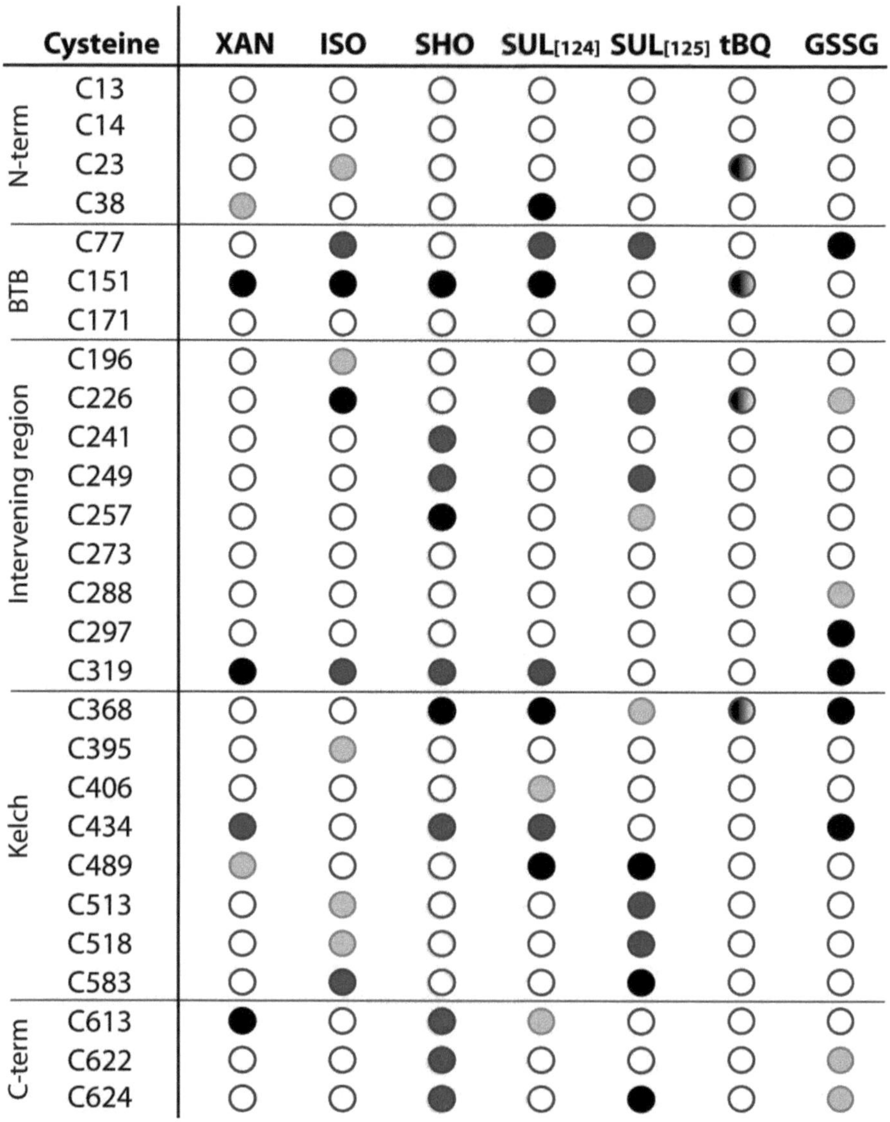

Fig. 7.5 Keap1 cysteines readily modified by phytochemicals, glutathione, and a quinone-forming polyphenol. Abbreviated name, chemical name and reference: XAN, xanthohumol [123]; ISO, isoliquiritigenin [123]; SHO, 10-shogaol [123]; SUL, sulforaphane [124, 125]; tBQ, tert-butylquinone [127]; and GSSG, oxidized glutathione [82]. Cysteines are ranked in order of most readily modified for all but tBQ, as determined by increasing concentrations of the electrophile, with the *darkest circles* being the most readily modified, and the *empty circle* indicating weakly modified or not modified cysteines. For tBQ, the cysteines are not ranked

cysteine has not yet been evaluated as a sensor for phytochemicals, and it has been shown not to be required to sense the soft cations, such as As^{3+} and Se^{4+} , but there is an indication that it may play a role in sensing the Zn^{2+} ion [114].

Notably, none of the phytochemicals evaluated in Fig. 7.5 modified C273 or C288 appreciably, although GSH modified C288 to some extent. These cysteines are considered to be potential sensors, as their mutation to serine renders Keap1 unable to repress Nrf2 [107, 122]. As these cysteines have been implicated in sensing more reactive Michael acceptors—alkenals, prostaglandins, and nitro-fatty acids [114–116]—they may serve to detect primarily these endogenous signaling agents, rather than more stable phytochemical electrophiles. Keap1 C226 and C613, which sense heavy metal salts and other soft cations [114], are readily modified by phytochemicals, with isoliquiritigenin and sulforaphane targeting the former and xanthohumol, 10-shogaol, and, to some extent, sulforaphane targeting the latter. However, as described above, mutation of C226 or C613 to serine had no effect on the ability of Keap1 to sense sulforaphane [114]. It is likely that the C151 residue present in these mutant proteins could be modified by sulforaphane, leading to a loss of Keap1 repression of Nrf2. Further work is required to ascertain whether modification of Keap1 cysteines other than C151 contributes to Nrf2 activation by phytochemicals.

7.3.4.2 Mechanism of Nrf2 Activation upon Modification of Keap1 Cysteines

The means by which Nrf2 escapes Keap1–Cul3-directed ubiquitination and degradation is still under active investigation. Disruption of the Keap1–Nrf2 interaction upon modification of Keap1 cysteines, allowing Nrf2 to escape Keap1, was originally proposed on the basis of experiments using recombinant proteins [126]. However, further investigations of a similar nature showed that the overall affinity of the Keap1–Nrf2 complex is maintained after modification of Keap1 at sensor cysteines, including C151 and C288, by ARE inducers including sulforaphane and isoliquiritigenin [128]. While a reduced affinity of Keap1 for Nrf2 after treatment of cells with tBHQ has also been observed by pull-down assays [129], other studies have reported that Keap1 and Nrf2 remain associated after treatment of cells with tBHQ [107, 130], sulforaphane [107], a synthetic triterpenoid derivative, dh404 [131], or quercetin [132]. Pull-down assays are a useful but imprecise means of determining relative affinities of protein–protein interactions, making it difficult to ascertain whether the observed perturbation of the Keap1–Nrf2 complex occurs in response to inducers in a cellular context. It is noteworthy that preventing Nrf2 degradation is sufficient to promote Nrf2 nuclear accumulation and ARE activation [133]. It has been hypothesized that the affinity of the weaker Nrf2 DLG site is decreased after modification of Keap1 cysteines including C151, C273 and/or C288, leading to decreased ubiquitination, without affecting substantially the overall stability of the complex [122].

Significantly, the stability of the Keap1–Cul3 interaction, also essential for ubiquitination, is reduced upon treatment with ARE inducers as observed by pull-down assays [112, 134]. These studies show that the reduced affinity is dependent on Keap1 C151, as mutation of C151 to serine largely abolished the effect. Modification of the C151 residue has clearly been shown to disrupt Keap1-mediated ubiquitination of Nrf2, as mutation of C151 to tryptophan had the same effect as electrophile treatment of cells [120]. The Keap1 C151W mutant was largely unable to mediate Nrf2 ubiquitination, leading to stabilization of Nrf2 with concurrent activation of ARE-regulated genes, and the mutant had reduced affinity for Cul3, as assessed by the pull-down assay. Remarkably, Nrf2 binding to Keap1 in this assay was unaffected by the tryptophan mutation. While other Keap1 cysteines do very likely participate as well in Nrf2 activation, the results with Keap1 C151W in cells, as well as in a zebrafish model [115], show that modification of Keap1 C151 alone is sufficient for signaling for Nrf2 activation.

It has been proposed that the negative charge of the C151 thiolate anion is essential for Nrf2 turnover, and that modification of C151 by an electrophile neutralizes the negative charge, leading to the signaling event [114]. However, a series of 13 mutations at position 151 showed that the size of the residues was the key determinant, rather than hydrophobicity or charge, with the largest residue, tryptophan, showing the largest effect [120]. In particular, asparagine, with a relatively small partial molar volume (i.e., size) and no capacity to carry a negative charge, rendered the corresponding Keap1 mutant effective in suppressing Nrf2. Therefore, the size of the residue at position 151 appears to be the major determining factor for Keap1's ability to repress Nrf2. Since all phytochemicals that were found to react with the C151 thiolate have partial molar volumes larger than that of tryptophan, they would all be large enough to trigger the signaling mechanism.

Alternatively, formation of disulfide bonds between Keap1 cysteines has been proposed to mediate loss of Nrf2 repression. Disulfide-linked Keap1 dimers have been observed in extracts of cells treated with the chalcone derivative *bis*(2-hydroxybenzylidene)acetone, which is a phenolic Michael acceptor [135]. In addition, treatment of cells with oxidizing agents (e.g., H_2O_2, nitric oxide, hypochlorous acid, or *S*-nitrosocysteine) leads to a Keap1 dimer linked through C151 [136]. Nonenzymatic oxidation of phenolic compounds will generate phenoxyl radicals of variable stability (Fig. 7.2), and it has been proposed that these reactive intermediates could mediate the formation of thiyl radicals, promoting the formation of a disulfide bond or bonds between Keap1 cysteines [78]. Furthermore, both ROS and GSSG could also participate in Keap1 disulfide bond formation (Fig. 7.2). In addition, a Keap1 dimer that was stable under reducing conditions was induced by treatment of cells with a quinoid-forming polyphenol, tBHQ [107]. Formation of this non-disulfide-linked dimer was dependent on C151. Radical-mediated hydrogen abstraction from surface tryptophans or tyrosines by subsequent cross-linking may account for the formation of the non-reducible dimer [137]. Further studies are required to determine, on a structural/molecular level, how direct modification of C151 and other Keap1 residues by ARE inducers or GSSG, or Keap1 dimer formation, leads to the loss of their ability to repress Nrf2.

7.3.4.3 Role of Protein Kinases in Nrf2 Activation by Phytochemicals

In addition to the direct Keap1-repression–inactivation mechanism discussed above, a number of kinases have been implicated in Nrf2 activation. The kinase pathways that currently appear to be the most important for Nrf2 activation by phytochemicals are the ones mediated by phosphatidylinositol 3-kinase (PI3K)/Akt and mitogen-activated protein kinases (MAPKs). The PI3K/Akt pathway mediates carnosol-induced expression of the cytoprotective enzyme HO-1 in rat pheochromocytoma PC12 cells [138]. In addition, EGCG, guggulsterone, and piceatannol activate expression of cytoprotective enzymes including HO-1 in the human mammary epithelial (MCF10A) cells in a PI3K/Akt-dependent manner [101, 139, 140]. EGCG activation of Nrf2 in BAECs is also PI3K/Akt dependent [89]. The Nrf2 activation by CDDO-Im is dependent upon PI3K/Akt activity in ARPE-19 retinal epithelial cells [141].

The mechanism by which Nrf2 is regulated by PI3K/Akt has been investigated in some detail. The GSK-3β kinase, which is inactivated upon phosphorylation by Akt1, phosphorylates and negatively regulates Nrf2 via two distinct mechanisms [142, 143]. The phosphorylation of Nrf2 by GSK-3β leads to both cytoplasmic localization of Nrf2 [142] and Keap1-independent ubiquitination by the SCF/β-TrCP/Cul1 complex, leading to Nrf2 degradation [143]. A key cysteine residue, C124 in phosphatase and tensin homolog (PTEN), a negative regulator of the PI3K/Akt axis [144], is directly modified by biotinylated CDDO in ARPE-19 cells [141], and the same cysteine is required for guggulsterone-induced Nrf2 accumulation in MCF10A cells [101]. Remaining to be demonstrated is whether the modification of PTEN C124 by electrophilic agents also inactivates GSK-3β, and whether this in turn leads to Nrf2 accumulation and Nrf2 nuclear localization.

While the inhibition of GSK-3β is implicated in the initial response to Nrf2 activators, the re-activation of GSK-3β has been proposed to occur several hours later, leading to eventual downregulation of the response by increasing Nrf2 nuclear export [145]. In this proposed mechanism, H_2O_2 treatment of cells activates GSK-3β, leading to subsequent activation of Fyn/Src kinases, which then phosphorylate Nrf2 Y568. Phosphorylation of Y568 enhances the interaction of Nrf2 with the Crm1 nuclear export protein. This mechanism has not yet been evaluated for phytochemical ARE inducers. However, these investigations were prompted by the observation that genistein, an ARE inducer, is a tyrosine kinase inhibitor. Whether inhibition of Nrf2 Y568 phosphorylation is in part or wholly responsible for the ARE induction by genistein or other natural phenols is unknown.

The role of MAPKs in activation of Nrf2 is somewhat controversial. A large number of studies with various phytochemicals, including sulforaphane, phenethyl isothiocyanate (PEITC), curcumin, quercetin, and EGCG, have implicated MAPKs in Nrf2 activation (reviewed in [146]). In order to assess the importance of this kinase family in Nrf2 signaling, MAPKs were overexpressed in HEK293T cells, and potential sites of MAPK-dependent phosphorylation on Nrf2 were identified by mass spectrometry. As a result, five phosphorylated residues were identified [146]. An Nrf2 mutant with all five sites mutated to alanine showed a moderate decrease in the transcriptional activity of Nrf2, concomitant with a slight reduction in its nuclear accumulation. However, the stability of the Nrf2 protein, which is primarily controlled by Keap1, was not affected. The general conclusion drawn from this work is that direct phosphorylation of Nrf2 by MAPKs has limited contribution to modulating Nrf2 activity [110, 146]. However, the Nrf2 protein with five mutated residues was only evaluated for blocked phosphorylation by overexpression of c-Jun N-terminal kinase (JNK2), rather than each MAPK relevant for Nrf2 activation [146]. Therefore, an additional site on Nrf2 beyond those five identified by mass spectrometry may in fact be a relevant target. Regardless of whether direct Nrf2 phosphorylation is the means by which MAPKs activate Nrf2, the sheer number of studies that implicate this pathway in Nrf2 activation by phytochemicals calls for further mechanistic investigation.

Activation of Nrf2 is also mediated by other kinases such as PERK [147, 148] and casein kinase 2 [149, 150]. However, to our knowledge, these pathways have not been evaluated for phytochemical activation of Nrf2. In addition, protein kinase C (PKC) isoforms are involved in the upregulation of cytoprotective genes by several phytochemicals. PKC-dependent ARE gene upregulation was observed for curcumin in HUH7 hepatoma cells and human monocytes [151, 152], piceatannol in BAECs [153], and epigallocatechin in human monocytic cells [154]. Nrf2 nuclear accumulation and ARE activation are also PKC-dependent for tBHQ in HepG2 and rat H4IIEC3 hepatoma cell lines [129, 155, 156], as well as nontumorigenic human keratinocytes (HaCaT cells) [157]. However, ARE activation by quercetin or kaempferol was not dependent on PKC in Hepa1c1c7 hepatoma cells [90], nor was ARE activation by diallyl trisulfide dependent on PKC in HepG2 cells [158]. Further investigations are required to ascertain under what circumstances a phytochemical might activate Nrf2 through PKC isoforms.

7.4 Summary and Future Directions

Nrf2 is emerging as a master control of cytoprotective mechanisms important for defense against environmental challenges and stress factors. The phytochemical activation of Nrf2 promises to be an important mechanism in the prevention and amelioration of a wide variety of human diseases. In addition, Nrf2 is under investigation as a drug target [159], and there is much to be learned from investigations of phytochemical ARE inducers with regard to what makes an effective agent with minimal toxicity. A large effort has been put toward understanding the mechanisms by which Nrf2 is activated, although phytochemicals are still relatively understudied compared to other compounds [160].

In general, a common trait of agents that activate Nrf2 effectively is the ability to react with thiols either directly or upon bioactivation (spontaneous or enzyme-catalyzed). Alternative mechanisms of Nrf2 de-repression involving induction of oxidative stress, for example, have also been discussed. The electrophilic and/or oxidative nature of Nrf2 activators means that a balance between toxic and protective effects is necessary for safe administration of these agents. This balance is described in conventional drug development as a "therapeutic window" or effective dosage range. Within this range, beneficial effects are exhibited at optimal strength without being compromised by adverse processes. Our primary recommendation for future studies in the phytochemical activation of Nrf2 is that, as with conventional drugs, efforts are put toward understanding how this dosage range is widened. This will enable identification or development of more effective agents for disease prevention and strategies for administering those agents.

Widening this dosage range does not appear to be a trivial exercise. The challenge lies in the fact that a strategy for increasing potency that merely relies on higher reactivity of these electrophiles is likely to enhance the extent of deleterious events. These may range from effect dilution through GSH conjugation followed by efflux to severe cytotoxicity associated with covalent modification of unintended targets. At the same time, reduction of toxicity through attenuation of electrophilicity would be expected to compromise potency. However, nature provides some important examples of how an effective therapeutic window can be achieved, which involves both partners in the modification reaction: sensor cysteines and naturally occurring cytoprotective agents.

The chemistry of the sensor cysteines plays an important role in ensuring advantageous biological effects of phytochemicals. The unique reactivity of soft thiolates of the sensor cysteines implicated in ARE induction, including Keap1 C151, ensures a high degree of "chemoselectivity" through the intrinsic preference for π-conjugated and thereby polarizable (soft) electrophilic centers, such as those found in effective phytochemical activators of Nrf2 (Fig. 7.1 and Sect. 7.3.3). Chemoselectivity is further ensured by the high reactivity of the sensor cysteines implicated in ARE induction, including Keap1 C151. What remains to be explored is whether yet further preference toward particular sensor cysteines could be provided by the remaining structural features of the covalent modifiers, such as overall size or presence of particular functionalities that could interact favorably with the protein environment surrounding the thiol.

Certain phytochemical ARE inducers, either relatively potent ones or those with low or no observable toxicity, also provide clues as to the features that widen the therapeutic window. Thus, preference for sensor thiolates can be enhanced by neighboring hydroxyl groups guiding a thiolate toward an attack on an electrophile. Alternatively, since sensor thiolates are so reactive, poor Michael acceptors could conjugate to them preferentially and, thereby, act as viable ARE inducers, while evading less reactive "off-target" nucleophiles. Examples of phytochemicals that are expected to be relatively unreactive Michael acceptors include those deactivated by alkene polysubstitution, bulky neighbors, and/or extended conjugation (Sect. 7.3.3). Remarkably, these deactivating factors are also expected to facilitate the reverse elimination process either by enhancing the stability of the electrophile, or by destabilizing the adduct. The reversibility of conjugates is emerging as an important feature for high potency and low toxicity. As discussed in Sect. 7.3.3, continuous deconjugation may simultaneously contribute to both toxicity reduction due to the instability of off-target conjugates and potency enhancement through efflux prevention. In addition, there is much to learn as to how modification of Keap1 cysteines leads to Nrf2 de-repression, for example, whether the size of modifiers influences the extent of the response, as suggested by a mutagenesis study [120].

Another factor important in the therapeutic window for phytochemicals is the ability to induce Nrf2 by a variety of pathways. Key sensor cysteines of phytochemical ARE inducers are being identified—Keap1 C151 and C124 of PTEN kinase—and other cysteines will likely emerge as important in sensing ARE inducers. Depending on the nature of the chemical and biological events associated with a given phytochemical in a given condition, different signaling "nodes" can be activated, all leading to Nrf2–ARE regulation. For example, the ultimate messenger molecule producing the Nrf2-dependent signal may not only be the phytochemical itself and/or its metabolite, but a concomitantly formed ROS/RNS or GSSG, as the GSH:GSSG ratio drops. These secondary signaling messengers target not only Keap1 and PTEN, but also numerous other kinases, including those involved in Nrf2 regulation. For example, various flavonoids including quercetin and EGCG can associate noncovalently with and modulate a number of protein kinases with high affinity (reviewed in [161]). There appears to be an intricate network of regulatory proteins that modulate Nrf2 activity, and multiple sensing steps can lead to the ability of Nrf2 upregulation to be tuned as is appropriate. Importantly, the presence of such a network implies that, if multiple nodes are activated by a single phytochemical, the convergence of the responses may lead to a more pronounced effect. This rationale also provides the basis for the unexpected effectiveness exhibited by mixtures of Nrf2 activators. Indeed, impressive synergism has been observed in a few studies conducted thus far, including for glucosinolate breakdown products [54] and a commercially available phytochemical supplement mixture [162]. Nrf2 activation, while clearly important, is certainly not the only mediator of the cytoprotective and disease-preventive attributes of phytochemicals (reviewed in [163]). Thus, phytochemicals that modulate a number of targets in disease prevention within their "therapeutic window" will likely be most effective in humans.

In summary, despite the multitude of cellular targets affected by phytochemicals, upregulation of Nrf2 is emerging as a key factor in their cytoprotective properties. In addition, while it is unlikely that a single phytochemical or even a plant source will emerge as a magic bullet for disease prevention or amelioration, we expect that unraveling the chemical and biological aspects of the action of phytochemicals may lead to unprecedented opportunities. These prospects could range from dietary/supplement recommendations to phytochemical "cocktails" specially formulated for synergistic effects and to nature-inspired synthetic molecules harnessing the most effective features of the plant-derived prototypes.

Acknowledgments We thank Young-Joon Surh, Mark Hannink, Wendy A. Peer, Andrew D. Mesecar, and Albena Dinkova-Kostova for their thoughtful comments on this Perspective. We are grateful to the National Institutes of Health for financial support through the R03 CA128095 grant.

References

1. Pashkow FJ (2011) Oxidative stress and inflammation in heart disease: do antioxidants have a role in treatment and/or prevention? Int J Inflam 2011:514623
2. Kasai H, Kawai K (2006) Oxidative DNA damage: mechanisms and significance in health and disease. Antioxid Redox Signal 8:981–983
3. Rappaport SM, Li H, Grigoryan H, Funk WE, Williams ER (2012) Adductomics: characterizing exposures to reactive electrophiles. Toxicol Lett 213:83–90
4. Malhotra D, Portales-Casamar E, Singh A, Srivastava S, Arenillas D, Happel C, Shyr C, Wakabayashi N, Kensler TW, Wasserman WW, Biswal S (2010) Global mapping of binding sites for Nrf2 identifies novel targets in cell survival response through ChIP-Seq profiling and network analysis. Nucleic Acids Res 38:5718–5734
5. Hayes JD, McMahon M (2009) NRF2 and KEAP1 mutations: permanent activation of an adaptive response in cancer. Trends Biochem Sci 34:176–188
6. Zhao Z, Chen Y, Wang J, Sternberg P, Freeman ML, Grossniklaus HE, Cai J (2011) Age-related retinopathy in NRF2-deficient mice. PLoS ONE 6:e19456
7. Chen P-C, Vargas MR, Pani AK, Smeyne RJ, Johnson DA, Kan YW, Johnson JA (2009) Nrf2-mediated neuroprotection in the MPTP mouse model of Parkinson's disease: Critical role for the astrocyte. Proc Natl Acad Sci U S A 106:2933–2938
8. Sussan TE, Rangasamy T, Blake DJ, Malhotra D, El-Haddad H, Bedja D, Yates MS, Kombairaju P, Yamamoto M, Liby KT, Sporn MB, Gabrielson KL, Champion HC, Tuder RM, Kensler TW, Biswal S (2009) Targeting Nrf2 with the triterpenoid CDDO—imidazolide attenuates cigarette smoke-induced emphysema and cardiac dysfunction in mice. Proc Natl Acad Sci U S A 106:250–255
9. Ungvari Z, Bagi Z, Feher A, Recchia FA, Sonntag WE, Pearson K, de Cabo R, Csiszar A (2010) Resveratrol confers endothelial protection via activation of the antioxidant transcription factor Nrf2. Am J Physiol Heart Circ Physiol 299:H18–24
10. Fahey JW, Haristoy X, Dolan PM, Kensler TW, Scholtus I, Stephenson KK, Talalay P, Lozniewski A (2002) Sulforaphane inhibits extracellular, intracellular, and antibiotic-resistant strains of Helicobacter pylori and prevents benzo[α]pyrene-induced stomach tumors. Proc Natl Acad Sci U S A 99:7610–7615
11. Becks L, Prince M, Burson H, Christophe C, Broadway M, Itoh K, Yamamoto M, Mathis M, Orchard E, Shi R, McLarty J, Pruitt K, Zhang S, Kleiner-Hancock H (2010) Aggressive mammary carcinoma progression in Nrf2 knockout mice treated with 7,12-dimethylbenz[a]anthracene. BMC Cancer 10:540
12. Xu C, Huang M-T, Shen G, Yuan X, Lin W, Khor TO, Conney AH, Kong A-NT (2006) Inhibition of 7,12-dimethylbenz(a)anthracene-induced skin tumorigenesis in C57BL/6 mice by sulforaphane is mediated by nuclear factor E2–related factor 2. Cancer Res 66:8293–8296
13. Zhao J, Moore AN, Redell JB, Dash PK (2007) Enhancing expression of Nrf2-driven genes protects the blood brain barrier after brain injury. J Neurosci 27:10240–10248
14. Aoki Y, Sato H, Nishimura N, Takahashi S, Itoh K, Yamamoto M (2001) Accelerated DNA adduct formation in the lung of the Nrf2 knockout mouse exposed to diesel exhaust. Toxicol Appl Pharmacol 173:154–160
15. Chan K, Kan YW (1999) Nrf2 is essential for protection against acute pulmonary injury in mice. Proc Natl Acad Sci U S A 96:12731–12736
16. Enomoto A, Itoh K, Nagayoshi E, Haruta J, Kimura T, O'Connor T, Harada T, Yamamoto M (2001) High sensitivity of Nrf2 knockout mice to acetaminophen hepatotoxicity associated with decreased expression of ARE-regulated drug metabolizing enzymes and antioxidant genes. Toxicol Sci 59:169–177
17. Braun S, Hanselmann C, Gassmann MG, auf dem Keller U, Born-Berclaz C, Chan K, Kan YW, Werner S (2002) Nrf2 transcription factor, a novel target of keratinocyte growth factor action which regulates gene expression and inflammation in the healing skin wound. Mol Cell Biol 22:5492–5505

18. Sykiotis GP, Bohmann D (2010) Stress-activated cap'n'collar transcription factors in aging and human disease. Sci Signal 3:re3
19. Prochaska HJ, Santamaria AB, Talalay P (1992) Rapid detection of inducers of enzymes that protect against carcinogens. Proc Natl Acad Sci U S A 89:2394–2398
20. Kinghorn AD, Su BN, Jang DS, Chang LC, Lee D, Gu JQ, Carcache-Blanco EJ, Pawlus AD, Lee SK, Park EJ, Cuendet M, Gills JJ, Bhat K, Park HS, Mata-Greenwood E, Song LL, Jang M, Pezzuto JM (2004) Natural inhibitors of carcinogenesis. Planta Med 70:691–705
21. Fahey JW, Zalcmann AT, Talalay P (2001) The chemical diversity and distribution of glucosinolates and isothiocyanates among plants. Phytochemistry 56:5–51
22. Hecht SS (2000) Inhibition of carcinogenesis by isothiocyanates. Drug Metab Rev 32:395–411
23. Burow M, Wittstock U (2009) Regulation and function of specifier proteins in plants. Phytochem Rev 8:87–99
24. Agerbirk NN, De Vos M, Kim JH, Jander G (2009) Indole glucosinolate breakdown and its biological effects. Phytochem Rev 8:101–120
25. Talalay P, De Long MJ, Prochaska HJ (1988) Identification of a common chemical signal regulating the induction of enzymes that protect against chemical carcinogenesis. Proc Natl Acad Sci U S A 85:8261–8265
26. Murakami A, Takahashi D, Kinoshita T, Koshimizu K, Kim HW, Yoshihiro A, Nakamura Y, Jiwajinda S, Terao J, Ohigashi H (2002) Zerumbone, a Southeast Asian ginger sesquiterpene, markedly suppresses free radical generation, proinflammatory protein production, and cancer cell proliferation accompanied by apoptosis: the α,β-unsaturated carbonyl group is a prerequisite. Carcinogenesis 23:795–802
27. Su BN, Jung Park E, Vigo JS, Graham JG, Cabieses F, Fong HH, Pezzuto JM, Kinghorn AD (2003) Activity-guided isolation of the chemical constituents of *Muntingia calabura* using a quinone reductase induction assay. Phytochemistry 63:335–341
28. Uda Y, Price KR, Williamson G, Rhodes MJ (1997) Induction of the anticarcinogenic marker enzyme, quinone reductase, in murine hepatoma cells in vitro by flavonoids. Cancer Lett 120:213–216
29. Su BN, Gu J-Q, Kang Y-H, Park E-J, Pezzuto JM, Kinghorn AD (2004) Induction of the phase II enzyme, quinone reductase, by withanolides and norwithanolides from Solanaceous species. Mini-Rev Org Chem 1:115–123
30. Dinkova-Kostova AT, Massiah MA, Bozak RE, Hicks RJ, Talalay P (2001) Potency of Michael reaction acceptors as inducers of enzymes that protect against carcinogenesis depends on their reactivity with sulfhydryl groups. Proc Natl Acad Sci U S A 98:3404–3409
31. Itoh K, Wakabayashi N, Katoh Y, Ishii T, Igarashi K, Engel JD, Yamamoto M (1999) Keap1 represses nuclear activation of antioxidant responsive elements by Nrf2 through binding to the amino-terminal Neh2 domain. Genes Dev 13:76–86
32. Prochaska HJ, Bregman HS, De Long MJ, Talalay P (1985) Specificity of induction of cancer protective enzymes by analogues of *tert*-butyl-4-hydroxyanisole (BHA). Biochem Pharmacol 34:3909–3914
33. Jacobs H, Moalin M, Bast A, van der Vijgh WJF, Haenen GRMM (2010) An essential difference between the flavonoids monoHER and quercetin in their interplay with the endogenous antioxidant network. PLoS ONE 5:e13880
34. Awad HM, Boersma MG, Boeren S, van Bladeren PJ, Vervoort J, Rietjens IMCM (2003) Quenching of quercetin quinone/quinone methides by different thiolate scavengers: stability and reversibility of conjugate formation. Chem Res Toxicol 16:822–831
35. Galati G, Moridani MY, Chan TS, O'Brien PJ (2001) Peroxidative metabolism of apigenin and naringenin versus luteolin and quercetin: glutathione oxidation and conjugation. Free Radic Biol Med 30:370–382
36. van Zanden JJ, Ben Hamman O, van Iersel MLPS, Boeren S, Cnubben NHP, Lo Bello M, Vervoort J, van Bladeren PJ, Rietjens IMCM (2003) Inhibition of human glutathione S-transferase P1–1 by the flavonoid quercetin. Chem Biol Interact 145:139–148

37. Ishii T, Mori T, Tanaka T, Mizuno D, Yamaji R, Kumazawa S, Nakayama T, Akagawa M (2008) Covalent modification of proteins by green tea polyphenol (-)-epigallocatechin-3-gallate through autoxidation. Free Radic Biol Med 45:1384–1394
38. Muzolf-Panek M, Gliszczyńska-Świgło A, de Haan L, Aarts JMMJG, Szymusiak H, Vervoort JM, Tyrakowska B, Rietjens IMCM (2008) Role of catechin quinones in the induction of EpRE-mediated gene expression. Chem Res Toxicol 21:2352–2360
39. Sang S, Lambert JD, Hong J, Tian S, Lee M-J, Stark RE, Ho C-T, Yang CS (2005) Synthesis and structure identification of thiol conjugates of (-)-epigallocatechin gallate and their urinary levels in mice. Chem Res Toxicol 18:1762–1769
40. Nakamura Y, Kumagai T, Yoshida C, Naito Y, Miyamoto M, Ohigashi H, Osawa T, Uchida K (2003) Pivotal role of electrophilicity in glutathione S-transferase induction by *tert*-butylhydroquinone. Biochemistry 42:4300–4309
41. Bensasson RV, Zoete V, Dinkova-Kostova AT, Talalay P (2008) Two-step mechanism of induction of the gene expression of a prototypic cancer-protective enzyme by diphenols. Chem Res Toxicol 21:805–812
42. Wang XJ, Hayes JD, Higgins LG, Wolf CR, Dinkova-Kostova AT (2010) Activation of the NRF2 signaling pathway by copper-mediated redox cycling of para- and ortho-hydroquinones. Chem Biol 17:75–85
43. Dinkova-Kostova AT, Wang XJ (2011) Induction of the Keap1/Nrf2/ARE pathway by oxidizable diphenols. Chem Biol Interact 192:101–106
44. Sparnins VL, Barany G, Wattenberg LW (1988) Effects of organosulfur compounds from garlic and onions on benzo[a]pyrene-induced neoplasia and glutathione S-transferase activity in the mouse. Carcinogenesis 9:131–134
45. Brady JF, Ishizaki H, Fukuto JM, Lin MC, Fadel A, Gapac JM, Yang CS (1991) Inhibition of cytochrome P-450 2E1 by diallyl sulfide and its metabolites. Chem Res Toxicol 4:642–647
46. Gurtoo HL, Marinello AJ, Struck RF, Paul B, Dahms RP (1981) Studies on the mechanism of denaturation of cytochrome P-450 by cyclophosphamide and its metabolites. J Biol Chem 256:11691–11701
47. Jin L, Baillie TA (1997) Metabolism of the chemoprotective agent diallyl sulfide to glutathione conjugates in rats. Chem Res Toxicol 10:318–327
48. Cavin C, Bezencon C, Guignard G, Schilter B (2003) Coffee diterpenes prevent benzo[a]pyrene genotoxicity in rat and human culture systems. Biochem Biophys Res Commun 306:488–495
49. Hwang YP, Jeong HG (2008) The coffee diterpene kahweol induces heme oxygenase-1 via the PI3K and p38/Nrf2 pathway to protect human dopaminergic neurons from 6-hydroxydopamine-derived oxidative stress. FEBS Lett 582:2655–2662
50. van Cruchten STJ, de Haan LHJ, Mulder PPJ, Kunne C, Boekschoten MV, Katan MB, Aarts JMMJG, Witkamp RF (2010) The role of epoxidation and electrophile-responsive element-regulated gene transcription in the potentially beneficial and harmful effects of the coffee components cafestol and kahweol. J Nutr Biochem 21:757–763
51. Boyd MR (1981) Toxicity mediated by reactive metabolites of furans. Adv Exp Med Biol 136 Pt B:865–879
52. Avonto C, Taglialatela-Scafati O, Pollastro F, Minassi A, Di Marzo V, De Petrocellis L, Appendino G (2011) An NMR spectroscopic method to identify and classify thiol-trapping agents: revival of Michael acceptors for drug discovery? Angewandte Chemie Int Ed 50:467–471
53. Khojasteh-Bakht SC, Chen W, Koenigs LL, Peter RM, Nelson SD (1999) Metabolism of (R)-(+)-pulegone and (R)-(+)-menthofuran by human liver cytochrome P-450s: evidence for formation of a furan epoxide. Drug Metab Dispos 27:574–580
54. Saw CL-L, Cintrón M, Wu T-Y, Guo Y, Huang Y, Jeong W-S, Kong A-NT (2011) Pharmacodynamics of dietary phytochemical indoles I3C and DIM: Induction of Nrf2-mediated phase II drug metabolizing and antioxidant genes and synergism with isothiocyanates. Biopharm Drug Dispos 32:289–300
55. Aggarwal BB, Ichikawa H (2005) Molecular targets and anticancer potential of indole-3-carbinol and its derivatives. Cell Cycle 4:1201–1215

56. Leibelt DA, Hedstrom OR, Fischer KA, Pereira CB, Williams DE (2003) Evaluation of chronic dietary exposure to indole-3-carbinol and absorption-enhanced 3,3-diindolylmethane in sprague-dawley rats. Toxicol Sci 74:10–21
57. Fordtran JS, Scroggie WB, Polter DE (1964) Colonic absorption of tryptophan metabolites in man. J Lab Clin Med 64:125–132
58. Thornton-Manning J, Appleton ML, Gonzalez FJ, Yost GS (1996) Metabolism of 3-methylindole by vaccinia-expressed P450 enzymes: correlation of 3-methyleneindolenine formation and protein-binding. J Pharmacol Exp Ther 276:21–29
59. Regal KA, Laws GM, Yuan C, Yost GS, Skiles GL (2001) Detection and characterization of DNA adducts of 3-methylindole. Chem Res Toxicol 14:1014–1024
60. Grose KR, Bjeldanes LF (1992) Oligomerization of indole-3-carbinol in aqueous acid. Chem Res Toxicol 5:188–193
61. Staub RE, Onisko B, Bjeldanes LF (2006) Fate of 3,3-diindolylmethane in cultured MCF-7 human breast cancer cells. Chem Res Toxicol 19:436–442
62. Sepkovic DW, Bradlow HL, Bell M (2001) Quantitative determination of 3,3-diindolylmethane in urine of individuals receiving indole-3-carbinol. Nutr Cancer 41:57–63
63. Ben-Dor A, Steiner M, Gheber L, Danilenko M, Dubi N, Linnewiel K, Zick A, Sharoni Y, Levy J (2005) Carotenoids activate the antioxidant response element transcription system. Mol Cancer Ther 4:177–186
64. Lewinsohn E, Sitrit Y, Bar E, Azulay Y, Meir A, Zamir D, Tadmor Y (2005) Carotenoid pigmentation affects the volatile composition of tomato and watermelon fruits, as revealed by comparative genetic analyses. J Agric Food Chem 53:3142–3148
65. Kondratyuk TP, Park E-J, Marler LE, Ahn S, Yuan Y, Choi Y, Yu R, van Breemen RB, Sun B, Hoshino J, Cushman M, Jermihov KC, Mesecar AD, Grubbs CJ, Pezzuto JM (2011) Resveratrol derivatives as promising chemopreventive agents with improved potency and selectivity. Mol Nutr Food Res 55:1249–1265
66. Kode A, Rajendrasozhan S, Caito S, Yang S-R, Megson IL, Rahman I (2008) Resveratrol induces glutathione synthesis by activation of Nrf2 and protects against cigarette smoke-mediated oxidative stress in human lung epithelial cells. Am J Physiol Lung Cell Mol Physiol 294:L478–L488
67. Piver B, Fer M, Vitrac X, Merillon JM, Dreano Y, Berthou F, Lucas D (2004) Involvement of cytochrome P450 1A2 in the biotransformation of trans-resveratrol in human liver microsomes. Biochem Pharmacol 68:773–782
68. Potter GA, Patterson LH, Wanogho E, Perry PJ, Butler PC, Ijaz T, Ruparelia KC, Lamb JH, Farmer PB, Stanley LA, Burke MD (2002) The cancer preventative agent resveratrol is converted to the anticancer agent piceatannol by the cytochrome P450 enzyme CYP1B1. Br J Cancer 86:774–778
69. Rushmore TH, Kong AN (2002) Pharmacogenomics, regulation and signaling pathways of phase I and II drug metabolizing enzymes. Curr Drug Metab 3:481–490
70. Casper RF, Quesne M, Rogers IM, Shirota T, Jolivet A, Milgrom E, Savouret J-F (1999) Resveratrol has antagonist activity on the aryl hydrocarbon receptor: implications for prevention of dioxin toxicity. Mol Pharmacol 56:784–790
71. Miao W, Hu L, Kandouz M, Batist G (2003) Oltipraz is a bifunctional inducer activating both phase I and phase II drug-metabolizing enzymes via the xenobiotic responsive element. Mol Pharmacol 64:346–354
72. Montano MM, Katzenellenbogen BS (1997) The quinone reductase gene: A unique estrogen receptor-regulated gene that is activated by antiestrogens. Proc Natl Acad Sci U S A 94:2581–2586
73. Montano MM, Jaiswal AK, Katzenellenbogen BS (1998) Transcriptional regulation of the human quinone reductase gene by antiestrogen-liganded estrogen receptor-alpha and estrogen receptor-beta. J Biol Chem 273:25443–25449
74. Montano MM, Deng H, Liu M, Sun X, Singal R (2004) Transcriptional regulation by the estrogen receptor of antioxidative stress enzymes and its functional implications. Oncogene 23:2442–2453

75. Bianco NR, Chaplin LJ, Montano MM (2005) Differential induction of quinone reductase by phytoestrogens and protection against oestrogen-induced DNA damage. Biochem J 385:279–287
76. Buryanovskyy L, Fu Y, Boyd M, Ma Y, Hsieh T-c, Wu JM, Zhang Z (2004) Crystal structure of quinone reductase 2 in complex with resveratrol. Biochemistry 43:11417–11426
77. Boutin JA, Chatelain-Egger F, Vella F, Delagrange P, Ferry G (2005) Quinone reductase 2 substrate specificity and inhibition pharmacology. Chem Biol Interact 151:213–228
78. Dinkova-Kostova AT, Talalay P (2008) Direct and indirect antioxidant properties of inducers of cytoprotective proteins. Mol Nutr Food Res 52:S128–S138
79. Nakamura Y, Miyoshi N (2010) Electrophiles in foods: the current status of isothiocyanates and their chemical biology. Biosci Biotechnol Biochem 74:242–255
80. Holland R, Navamal M, Velayutham M, Zweier JL, Kensler TW, Fishbein JC (2009) Hydrogen peroxide is a second messenger in phase 2 enzyme induction by cancer chemopreventive dithiolethiones. Chem Res Toxicol 22:1427–1434
81. Winterbourn CC (1993) Superoxide as an intracellular radical sink. Free Radic Biol Med 14:85–90
82. Holland R, Hawkins AE, Eggler AL, Mesecar AD, Fabris D, Fishbein JC (2008) Prospective type 1 and type 2 disulfides of Keap1 protein. Chem Res Toxicol 21:2051–2060
83. Mieyal JJ, Gallogly MM, Qanungo S, Sabens EA, Shelton MD (2008) Molecular mechanisms and clinical implications of reversible protein S-glutathionylation. Antioxid Redox Signal 10:1941–1988
84. Na H-K, Surh Y-J (2006) Transcriptional regulation via cysteine thiol modification: A novel molecular strategy for chemoprevention and cytoprotection. Mol Carcinog 45:368–380
85. Janssen-Heininger YMW, Mossman BT, Heintz NH, Forman HJ, Kalyanaraman B, Finkel T, Stamler JS, Rhee SG, van der Vliet A (2008) Redox-based regulation of signal transduction: Principles, pitfalls, and promises. Free Radic Biol Med 45:1–17
86. Nakagawa H, Hasumi K, Woo J-T, Nagai K, Wachi M (2004) Generation of hydrogen peroxide primarily contributes to the induction of Fe(II)-dependent apoptosis in Jurkat cells by (-)-epigallocatechin gallate. Carcinogenesis 25:1567–1574
87. Yang G-Y, Liao J, Li C, Chung J, Yurkow EJ, Ho C-T, Yang CS (2000) Effect of black and green tea polyphenols on c-jun phosphorylation and H2O2 production in transformed and non-transformed human bronchial cell lines: possible mechanisms of cell growth inhibition and apoptosis induction. Carcinogenesis 21:2035–2039
88. Hou Z, Sang S, You H, Lee M-J, Hong J, Chin K-V, Yang CS (2005) Mechanism of action of (-)-epigallocatechin-3-gallate: auto-oxidation-dependent inactivation of epidermal growth factor receptor and direct effects on growth inhibition in human esophageal cancer KYSE 150 cells. Cancer Res 65:8049–8056
89. Wu CC, Hsu MC, Hsieh CW, Lin JB, Lai PH, Wung BS (2006) Upregulation of heme oxygenase-1 by Epigallocatechin-3-gallate via the phosphatidylinositol 3-kinase/Akt and ERK pathways. Life Sci 78:2889–2897
90. Lee-Hilz YY, Boerboom A-MJF, Westphal AH, van Berkel WJH, Aarts JMMJG, Rietjens IMCM (2006) Pro-oxidant activity of flavonoids induces EpRE-mediated gene expression. Chem Res Toxicol 19:1499–1505
91. Galati G, Moridani MY, Chan TS, O‚ÄôBrien PJ (2001) Peroxidative metabolism of apigenin and naringenin versus luteolin and quercetin: glutathione oxidation and conjugation. Free Radic Biol Med 30:370–382
92. Galati G, Chan T, Wu B, O'Brien PJ (1999) Glutathione-dependent generation of reactive oxygen species by the peroxidase-catalyzed redox cycling of flavonoids. Chem Res Toxicol 12:521–525
93. Liby KT, Yore MM, Sporn MB (2007) Triterpenoids and rexinoids as multifunctional agents for the prevention and treatment of cancer. Nat Rev Cancer 7:357–369
94. Forman HJ, Fukuto JM, Torres M (2004) Redox signaling: thiol chemistry defines which reactive oxygen and nitrogen species can act as second messengers. Am J Physiol Cell Physiol 287:C246–256
95. Magder S (2006) Reactive oxygen species: toxic molecules or spark of life? Crit Care 10:208

96. Liby K, Hock T, Yore MM, Suh N, Place AE, Risingsong R, Williams CR, Royce DB, Honda T, Honda Y, Gribble GW, Hill-Kapturczak N, Agarwal A, Sporn MB (2005) The synthetic triterpenoids, CDDO and CDDO-imidazolide, are potent inducers of heme oxygenase-1 and Nrf2/ARE signaling. Cancer Res 65:4789–4798
97. Calabrese EJ (2005) Cancer biology and hormesis: human tumor cell lines commonly display hormetic (biphasic) dose responses. Crit Rev Toxicol 35:463–582
98. LoPachin RM, Barber DS, Gavin T (2008) Molecular mechanisms of the conjugated alpha, beta-unsaturated carbonyl derivatives: relevance to neurotoxicity and neurodegenerative diseases. Toxicol Sci 104:235–249
99. Brandes N, Schmitt S, Jakob U (2009) Thiol-based redox switches in eukaryotic proteins. Antioxid Redox Signal 11:997–1014
100. Sporn MB, Liby KT, Yore MM, Fu L, Lopchuk JM, Gribble GW (2011) New synthetic triterpenoids: potent agents for prevention and treatment of tissue injury caused by inflammatory and oxidative stress. J Nat Prod 74:537–545
101. Almazari I, Park J-M, Park S-A, Suh J-Y, Na H-K, Cha Y-N, Surh Y-J (2012) Guggulsterone induces heme oxygenase-1 expression through activation of Nrf2 in human mammary epithelial cells: PTEN as a putative target. Carcinogenesis 33:368–376
102. Lin D, Saleh S, Liebler DC (2008) Reversibility of covalent electrophile-protein adducts and chemical toxicity. Chem Res Toxicol 21:2361–2369
103. Hayes JD, Flanagan JU, Jowsey IR (2005) Glutathione transferases. Annu Rev Pharmacol Toxicol 45:51–88
104. Talalay P, Fahey JW, Holtzclaw WD, Prestera T, Zhang Y (1995) Chemoprotection against cancer by phase 2 enzyme induction. Toxicol Lett 82–83:173–179
105. Thornalley PJ (2002) Isothiocyanates: mechanism of cancer chemopreventive action. Anti-cancer Drugs 13:331–338
106. Klaić L, Morimoto RI, Silverman RB (2012) Celastrol analogues as inducers of the heat shock response. Design and synthesis of affinity probes for the identification of protein targets. ACS Chem Biol 7:928–937
107. Zhang DD, Hannink M (2003) Distinct cysteine residues in Keap1 are required for Keap1-dependent ubiquitination of Nrf2 and for stabilization of Nrf2 by chemopreventive agents and oxidative stress. Mol Cell Biol 23:8137–8151
108. Karapetian RN, Evstafieva AG, Abaeva IS, Chichkova NV, Filonov GS, Rubtsov YP, Sukhacheva EA, Melnikov SV, Schneider U, Wanker EE, Vartapetian AB (2005) Nuclear oncoprotein prothymosin alpha is a partner of Keap1: Implications for expression of oxidative stress-protecting genes. Mol Cell Biol 25:1089–1099
109. Velichkova M, Hasson T (2005) Keap1 regulates the oxidation-sensitive shuttling of Nrf2 into and out of the nucleus via a Crm1-dependent nuclear export mechanism. Mol Cell Biol 25:4501–4513
110. Hayes JD, McMahon M, Chowdhry S, Dinkova-Kostova AT (2010) Cancer chemoprevention mechanisms mediated through the Keap1-Nrf2 pathway. Antioxid Redox Signal 13:1713–1748
111. Tong KI, Katoh Y, Kusunoki H, Itoh K, Tanaka T, Yamamoto M (2006) Keap1 recruits Neh2 through binding to ETGE and DLG motifs: characterization of the two-site molecular recognition model. Mol Cell Biol 26:2887–2900
112. Zhang DD, Lo S-C, Cross JV, Templeton DJ, Hannink M (2004) Keap1 is a redox-regulated substrate adaptor protein for a Cul3-dependent ubiquitin ligase complex. Mol Cell Biol 24:10941–10953
113. McMahon M, Thomas N, Itoh K, Yamamoto M, Hayes JD (2006) Dimerisation of substrate adaptors can facilitate cullin-mediated ubiquitylation of proteins by a 'tethering' mechanism: A two-site interaction model for the Nrf2-Keap1 complex. J Biol Chem 281:24756–24768
114. McMahon M, Lamont DJ, Beattie KA, Hayes JD (2010) Keap1 perceives stress via three sensors for the endogenous signaling molecules nitric oxide, zinc, and alkenals. Proc Natl Acad Sci U S A 107:18838–18843

115. Kobayashi M, Li L, Iwamoto N, Nakajima-Takagi Y, Kaneko H, Nakayama Y, Eguchi M, Wada Y, Kumagai Y, Yamamoto M (2009) The antioxidant defense system Keap1-Nrf2 comprises a multiple sensing mechanism for responding to a wide range of chemical compounds. Mol Cell Biol 29:493–502
116. Kansanen E, Bonacci G, Schopfer FJ, Kuosmanen SM, Tong KI, Leinonen H, Woodcock SR, Yamamoto M, Carlberg C, Ylä-Herttuala S, Freeman BA, Levonen A-L (2011) Electrophilic nitro-fatty acids activate NRF2 by a KEAP1 cysteine 151-independent mechanism. J Biol Chem 286:14019–14027
117. Sakurai T, Kanayama M, Shibata T, Itoh K, Kobayashi A, Yamamoto M, Uchida K (2006) Ebselen, a seleno-organic antioxidant, as an electrophile. Chem Res Toxicol 19:1196–1204
118. Ohnuma T, Nakayama S, Anan E, Nishiyama T, Ogura K, Hiratsuka A (2010) Activation of the Nrf2/ARE pathway via S-alkylation of cysteine 151 in the chemopreventive agent-sensor Keap1 protein by falcarindiol, a conjugated diacetylene compound. Toxicol Appl Pharmacol 244:27–36
119. Satoh T, Okamoto S-I, Cui J, Watanabe Y, Furuta K, Suzuki M, Tohyama K, Lipton SA (2006) From the cover: activation of the Keap1/Nrf2 pathway for neuroprotection by electrophillic phase II inducers. Proc Natl Acad Sci U S A 103:768–773
120. Eggler AL, Small E, Hannink M, Mesecar AD (2009) Cul3-mediated Nrf2 ubiquitination and antioxidant response element (ARE) activation are dependent on the partial molar volume at position 151 of Keap1. Biochem J 422:171–180
121. Hur W, Sun Z, Jiang T, Mason DE, Peters EC, Zhang DD, Luesch H, Schultz PG, Gray NS (2010) A small-molecule inducer of the antioxidant response element. Chem Biol 17:537–547
122. Yamamoto T, Suzuki T, Kobayashi A, Wakabayashi J, Maher J, Motohashi H, Yamamoto M (2008) Physiological significance of reactive cysteine residues of Keap1 in determining Nrf2 activity. Mol Cell Biol 28:2758–2770
123. Luo Y, Eggler AL, Liu D, Liu G, Mesecar AD, van Breemen RB (2007) Sites of alkylation of human Keap1 by natural chemoprevention agents. J Am Soc Mass Spectrom 18:2226–2232
124. Hu C, Eggler AL, Mesecar AD, van Breemen RB (2011) Modification of Keap1 cysteine residues by sulforaphane. Chem Res Toxicol 24:515–521
125. Hong F, Freeman ML, Liebler DC (2005) Identification of sensor cysteines in human Keap1 modified by the cancer chemopreventive agent sulforaphane. Chem Res Toxicol 18:1917–1926
126. Dinkova-Kostova AT, Holtzclaw WD, Cole RN, Itoh K, Wakabayashi N, Katoh Y, Yamamoto M, Talalay P (2002) Direct evidence that sulfhydryl groups of Keap1 are the sensors regulating induction of phase 2 enzymes that protect against carcinogens and oxidants. Proc Natl Acad Sci U S A 99:11908–11913
127. Abiko Y, Miura T, Phuc BH, Shinkai Y, Kumagai Y (2011) Participation of covalent modification of Keap1 in the activation of Nrf2 by tert-butylbenzoquinone, an electrophilic metabolite of butylated hydroxyanisole. Toxicol Appl Pharmacol 255:32–39
128. Eggler AL, Liu G, Pezzuto JM, van Breemen RB, Mesecar AD (2005) Modifying specific cysteines of the electrophile-sensing human Keap1 protein is insufficient to disrupt binding to the Nrf2 domain Neh2. Proc Natl Acad Sci U S A 102:10070–10075
129. Niture SK, Jain AK, Jaiswal AK (2009) Antioxidant-induced modification of INrf2 cysteine 151 and PKC-delta-mediated phosphorylation of Nrf2 serine 40 are both required for stabilization and nuclear translocation of Nrf2 and increased drug resistance. J Cell Sci 122:4452–4464
130. Kobayashi A, Kang M-I, Watai Y, Tong KI, Shibata T, Uchida K, Yamamoto M (2006) Oxidative and electrophilic stresses activate Nrf2 through inhibition of ubiquitination activity of Keap1. Mol Cell Biol 26:221–229
131. Ichikawa T, Li J, Meyer CJ, Janicki JS, Hannink M, Cui T (2009) Dihydro-CDDO-trifluoroethyl amide (dh404), a novel Nrf2 activator, suppresses oxidative stress in cardiomyocytes. PLoS ONE 4:e8391
132. Tanigawa S, Fujii M, Hou D-X (2007) Action of Nrf2 and Keap1 in ARE-mediated NQO1 expression by quercetin. Free Radic Biol Med 42:1690–1703

133. Dreger H, Westphal K, Wilck N, Baumann G, Stangl V, Stangl K, Meiners S (2010) Protection of vascular cells from oxidative stress by proteasome inhibition depends on Nrf2. Cardiovasc Res 85:395–403
134. Rachakonda G, Xiong Y, Sekhar KR, Stamer SL, Liebler DC, Freeman ML (2008) Covalent modification at Cys151 dissociates the electrophile sensor Keap1 from the ubiquitin ligase CUL3. Chem Res Toxicol 21:705–710
135. Wakabayashi N, Dinkova-Kostova AT, Holtzclaw WD, Kang MI, Kobayashi A, Yamamoto M, Kensler TW, Talalay P (2004) Protection against electrophile and oxidant stress by induction of the phase 2 response: fate of cysteines of the Keap1 sensor modified by inducers. Proc Natl Acad Sci U S A 101:2040–2045
136. Fourquet S, Guerois R, Biard D, Toledano MB (2010) Activation of NRF2 by nitrosative agents and H2O2 involves KEAP1 disulfide formation. J Biol Chem 285:8463–8471
137. Lund MN, Luxford C, Skibsted LH, Davies MJ (2008) Oxidation of myosin by haem proteins generates myosin radicals and protein cross-links. Biochem J 410:565–574
138. Martin D, Rojo AI, Salinas M, Diaz R, Gallardo G, Alam J, De Galarreta CM, Cuadrado A (2004) Regulation of heme oxygenase-1 expression through the phosphatidylinositol 3-kinase/Akt pathway and the Nrf2 transcription factor in response to the antioxidant phytochemical carnosol. J Biol Chem 279:8919–8929
139. Na H-K, Kim E-H, Jung J-H, Lee H-H, Hyun J-W, Surh Y-J (2008) (-)-Epigallocatechin gallate induces Nrf2-mediated antioxidant enzyme expression via activation of PI3K and ERK in human mammary epithelial cells. Arch Biochem Biophys 476:171–177
140. Lee H-H, Park S-A, Almazari I, Kim E-H, Na H-K, Surh Y-J (2010) Piceatannol induces heme oxygenase-1 expression in human mammary epithelial cells through activation of ARE-driven Nrf2 signaling. Arch Biochem Biophys 501:142–150
141. Pitha-Rowe I, Liby K, Royce D, Sporn M (2009) Synthetic triterpenoids attenuate cytotoxic retinal injury: cross-talk between Nrf2 and PI3K/AKT signaling through inhibition of the lipid phosphatase PTEN. Invest Ophthalmol Vis Sci 50:5339–5347
142. Salazar M, Rojo AI, Velasco D, de Sagarra RM, Cuadrado A (2006) Glycogen synthase kinase-3beta inhibits the xenobiotic and antioxidant cell response by direct phosphorylation and nuclear exclusion of the transcription factor Nrf2. J Biol Chem 281:14841–14851
143. Rada P, Rojo AI, Chowdhry S, McMahon M, Hayes JD, Cuadrado A (2011) SCF/β-TrCP promotes glycogen synthase kinase 3-dependent degradation of the Nrf2 transcription factor in a Keap1-independent manner. Mol Cell Biol 31:1121–1133
144. Gericke A, Munson M, Ross AH (2006) Regulation of the PTEN phosphatase. Gene 374:1–9
145. Jain AK, Jaiswal AK (2007) GSK-3beta acts upstream of Fyn kinase in regulation of nuclear export and degradation of NF-E2 related factor 2. J Biol Chem 282:16502–16510
146. Sun Z, Huang Z, Zhang DD (2009) Phosphorylation of Nrf2 at multiple sites by MAP kinases has a limited contribution in modulating the Nrf2-dependent antioxidant response. PLoS ONE 4:e6588
147. Cullinan SB, Zhang D, Hannink M, Arvisais E, Kaufman RJ, Diehl JA (2003) Nrf2 is a direct PERK substrate and effector of PERK-dependent cell survival. Mol Cell Biol 23:7198–7209
148. Cullinan SB, Diehl JA (2006) Coordination of ER and oxidative stress signaling: The PERK/Nrf2 signaling pathway. Int J Biochem Cell Biol 38:317–332
149. Pi J, Bai Y, Reece JM, Williams J, Liu D, Freeman ML, Fahl WE, Shugar D, Liu J, Qu W, Collins S, Waalkes MP (2007) Molecular mechanism of human Nrf2 activation and degradation: Role of sequential phosphorylation by protein kinase CK2. Free Radic Biol Med 42:1797–1806
150. Apopa PL, He X, Ma Q (2008) Phosphorylation of Nrf2 in the transcription activation domain by casein kinase 2 (CK2) is critical for the nuclear translocation and transcription activation function of Nrf2 in IMR-32 neuroblastoma cells. J Biochem Mol Toxicol 22:63–76
151. McNally SJ, Harrison EM, Ross JA, Garden OJ, Wigmore SJ (2007) Curcumin induces heme oxygenase 1 through generation of reactive oxygen species, p38 activation and phosphatase inhibition. Int J Mol Med 19:165–172

152. Rushworth SA, Ogborne RM, Charalambos CA, O'Connell MA (2006) Role of protein kinase C δ in curcumin-induced antioxidant response element-mediated gene expression in human monocytes. Biochem Biophys Res Commun 341:1007–1016
153. Wung B-S, Hsu M-C, Wu C-C, Hsieh C-W (2006) Piceatannol upregulates endothelial heme oxygenase-1 expression via novel protein kinase C and tyrosine kinase pathways. Pharmacol Res 53:113–122
154. Ogborne RM, Rushworth SA, O'Connell MA (2008) Epigallocatechin activates haem oxygenase-1 expression via protein kinase Cdelta and Nrf2. Biochem Biophys Res Commun 373:584–588
155. Numazawa S, Ishikawa M, Yoshida A, Tanaka S, Yoshida T (2003) Atypical protein kinase C mediates activation of NF-E2-related factor 2 in response to oxidative stress. Am J Physiol Cell Physiol 285:C334–C342
156. Sherratt PJ, Huang HC, Nguyen T, Pickett CB (2004) Role of protein phosphorylation in the regulation of NF-E2-related factor 2 activity. Methods Enzymol 378:286–301
157. Zhu M, Zhang Y, Bowden GT (2006) Involvement of mitogen-activated protein kinases and protein kinase C in regulation of antioxidant response element activity in human keratinocytes. Cancer Lett 244:220–228
158. Chen C, Pung D, Leong V, Hebbar V, Shen G, Nair S, Li W, Kong AN (2004) Induction of detoxifying enzymes by garlic organosulfur compounds through transcription factor Nrf2: effect of chemical structure and stress signals. Free Radic Biol Med 37:1578–1590
159. Kern JT, Hannink M, Hess JF (2007) Disruption of the Keap1-containing ubiquitination complex as an antioxidant therapy. Curr Top Med Chem 7:972–978
160. Eggler AL, Gay KA, Mesecar AD (2008) Molecular mechanisms of natural products in chemoprevention: induction of cytoprotective enzymes by Nrf2. Mol Nutr Food Res 52(Suppl 1):S84–S94
161. Hou DX, Kumamoto T (2010) Flavonoids as protein kinase inhibitors for cancer chemoprevention: direct binding and molecular modeling. Antioxid Redox Signal 13:691–719
162. Velmurugan K, Alam J, McCord JM, Pugazhenthi S (2009) Synergistic induction of heme oxygenase-1 by the components of the antioxidant supplement Protandim. Free Radic Biol Med 46:430–440
163. Murakami A, Ohnishi K (2012) Target molecules of food phytochemicals: Food science bound for the next dimension. Food Funct 3:462–476

Index

Symbols
10-shogaol, 139, 141

A
Abiotic stress, 3, 60
Allelopathy, 88
Anthraquinone, 46, 85
Antioxidant, 22, 34, 97, 101, 123, 133, 137

B
Biosynthesis, 16
  avenanthramide, 63, 65
  camptothecin, 49–51
  chlorophyll, 7
  dihydro-analogues, 15
  hemiterpenoid glycosides, 17
  HTG, 7
  isoprenoid, 13
  pathogenesis related proteins, 57
  phtoalexin, 63
  PR-1, 61
  prenyl quinones and tocopherols, 82
  salicylic acid, 58
  sorgoleone, 86–88
Biotransformation, 126
  diindolylmethane, 131
  indole-3-carbinol, 132

C
Callicarpenal, 74
Camptotheca acuminata, 44
Camptothecin, 44
  anti-tumor activity, 44
  biosynthesis, 49–51
  producing plants, in vitro cultures of, 46, 47
Celastrol, 100–103, 107
Cell and tissue culture, 46
Cereal grains, 55
Crop yield, 55
Curcumin, 104, 105, 107, 126, 143, 144

D
Dietary supplements, 22, 28, 38
  bioactive flavonoids in, 28
  proanthocyanidins, 37
Disease resistance, 56
  Flor's gene-for-gene theory, 57
  rice in field trials, 62

E
Electrophiles, 101, 109, 131, 135, 136, 145
  conjugation-stabilized, 134
  phytochemical, 141
  phytochemical-derieved, 132
  thiolate-reactive, 133
Essential oil, 73, 75, 76
  Cupressus funebris, 75
  Juniperus chinensis, 75
  Juniperus communis, 75
  Ligusticum mutellina, 75
  Pimpinella, 76

F
Food security, 56

G
Gedunin, 102, 103
Glutathione, 136, 137
  biosynthesis enzymes, 123
  oxidized, 139
  regeneration of, 97, 98

D. R. Gang (ed.), *50 Years of Phytochemistry Research*,
Recent Advances in Phytochemistry 43, DOI 10.1007/978-3-319-00581-2

H
Hairy roots, 50
  Ophiorrhiza species, 46
  RNAi, 51
Heat shock factor 1 (HSF1), 96, 98
Hemiterpenoids, 5, 10, 16, 17
Hesperidin, 28
  protonated ions, 28
  sodiated molecular ions, 30

I
Induced systemic resistance (ISR), 59
Insect repellent, 74, 75
Intermediol, 74
Ion mobility spectrometry (IMS), 22
Isoliquiritigenin, 139, 141
Isothiocyanates, 96, 124, 126, 135
  transthiocarbamoylation reactions, 107
Isoxanthohumol
  isobaric prenylated flavonoid, 34

K
KEAP1/NRF2/ARE pathway, 96–98
  activation of, 106–111
  induction of, 99
Kelch-like ECH-associated protein 1 (KEAP1), 96, 107, 109
  BTB domain, modeling of, 139
  Crm1-dependent nuclear export signal sequence, 138
  cysteine residues, 107, 108
  cysteines, 134
  dimer formation, 142
  disulfide bond formation, 142
  repressor function, 108
  secondary signaling messengers, 146

L
Leptospermone, 81–83
  biosynthetic pathway, 84
  bleaching of plant tissues, 83
  in vitro chemical synthesis, 83

M
Mass spectrometry (MS), 2, 138, 143, 144
  analyzers, 23
  bioactive flavonoids, analysis of, 28
Metabolomics, 1, 13, 16, 34, 38
  plant metabolome analysis, 2, 3
Methylerythritol phosphate (MEP) pathway, 15–17, 51
Michael acceptors, 104, 127, 141, 145
Monoterpenoid indole alkaloid, 49
  camptothecin, 44

N
Nuclear factor-erythroid 2 p45-related factor 2 (NRF2), 96, 97
  consequences of activation, 100–106
  phytochemical activation, 124, 126, 127, 129–139, 141–144
  phytochemicals involved in activation, 100–106
  upregulation of cytoprotective genes, 123
Nuclear factor-erythroid 2 p45-related factor 2 (NRF2) transcription factor, 123
  negative regulator of, 96
Nuclear magnetic resonance (NMR), 2
  dataset, 4
  high field, 2

O
Ophiorrhiza, 45, 46, 49

P
Pesticides, 64, 65, 72, 73
  arthropod, 73
Phytochemicals, 78, 122, 135, 138, 144
  bioactive, 23
  biological effects, 145
  characterization of, 35
  cytoprotective, 124, 126
  disease-preventive attributes, 146
  insecticidal, 73
  NRF2 activation, 100–106
  pesticidal, 73
  sulfhydryl-reactive, 111
Plant defense activators (PDA), 56
  commercial, 61, 62
  prospects for cereal crops, 65, 66
Plant-derived natural compounds
  callicarpenal, 74
  intermediol, 74
Proanthocyanidins, 27
  characterization of, 37
  dietary supplements, 37
  ESI-MS analysis, 24
  grape seed, 23
Pro-oxidant, 133

R
Reactive nitrogen species (RNS), 123
Reactive oxygen species (ROS), 123
  role in NRF2 activation, 133, 134
Rutin, 28, 30

S
Scopolin, 7, 17
Sorgoleone, 85
  biosynthesis, 86–88
  biosynthetic pathway, 86
Strictosidine synthase (STR), 49
Sulfhydryl-reactive phytochemicals, 111
Sulforaphane, 105–109, 124, 126, 127, 137–139, 141, 143
Systemic acquired resistance (SAR), 58, 59, 62, 63, 66
  biomarkers, 60
  studies, 60

T
Traveling wave ion mobility spectrometry mass spectrometry (TWIMS-MS), 22
  applications, 23
  biopolymers analysis, 24, 25, 27
  plant metabolite characterization, 34, 35, 37
Tryptophan decarboxylase (TDC), 49

W
Withaferin A, 104

X
Xanthohumol (XN), 34, 139, 141

MIX
Papier aus verantwortungsvollen Quellen
Paper from responsible sources
FSC® C105338

If you have any concerns about our products,
you can contact us on
ProductSafety@springernature.com

In case Publisher is established outside the EU,
the EU authorized representative is:
Springer Nature Customer Service Center GmbH
Europaplatz 3, 69115 Heidelberg, Germany

Printed by Libri Plureos GmbH
in Hamburg, Germany